MIDDLE GRADES MATHEMATICS PROJECT

Similarity and Equivalent Fractions

Glenda Lappan

William Fitzgerald

Mary Jean Winter

Elizabeth Phillips

Addison-Wesley Publishing Company

Menlo Park, California • Reading, Massachusetts • Don Mills, Ontario
Wokingham, England • Amsterdam • Sydney • Singapore
Tokyo • Mexico City • Bogotá • Santiago • San Juan

Acknowledgments

"Distortions: An Activity for Practice and Exploration" by Michael T. Battista from *Arithmetic Teacher*, January 1982. Reprinted by permission of the National Council of Teachers of Mathematics.

Pac-Man® is the licensed TM and © 1980 Bally Midway Mfg. Co. All rights reserved.

This book is published by the Addison-Wesley Innovative Division.

The blackline masters in this publication are designed to be used with appropriate duplicating equipment to reproduce copies for classroom use. Addison-Wesley Publishing Company grants permission to classroom teachers to reproduce these masters.

ISBN 0-201-21476-8

12 13 14 15 - ML - 95 94 93

About the authors

William Fitzgerald, Ph.D. in mathematics education, University of Michigan, joined the faculty of Michigan State University in 1966 and has been Professor of Mathematics and Education since 1971. He has had extensive experience at all levels of mathematics teaching and has been involved in the development of mathematics laboratories.

Glenda Lappan, B.A., Mercer University, Macon, Georgia, and Ed.D., University of Georgia, is Professor of Mathematics at Michigan State University. She directed the mathematics component of MSU Sloan Foundation Minority Engineering Project. She has taught high school mathematics and since 1976 has worked regularly with students and teachers of grades 3–8.

Elizabeth Phillips, B.S. in mathematics and chemistry, Wisconsin State University, and M.S. in mathematics, University of Notre Dame, was visiting scholar in mathematics education at Cambridge University, England. She conducts inservice workshops for teachers and is the author of several papers and books. Currently she is a faculty member in the Department of Mathematics at Michigan State University.

Janet Shroyer, B.S., Knox College, and Ph.D., Michigan State University, has taught mathematics in Lansing public schools and at Oregon College of Education. She was a consultant in the Office of Research Services, evaluator of a teacher corps project, and a research intern in the Institute for Research on Teaching. Presently she is Associate Professor in the Mathematics Department of Aquinas College, Grand Rapids, Michigan.

Mary Jean Winter, A.B., Vassar College, and Ph.D. in mathematics, Carnegie Institute of Technology, has been Professor of Mathematics at Michigan State University since 1965. She has been involved in mathematics education at both school and college (teacher training) level since 1975. She has been especially interested in developing middle school and secondary activities using computers and other manipulatives.

A special note of recognition

Sincere appreciation is expressed to the following persons for their significant contribution to the Middle Grades Mathematics Project.

Assistants:	**David Ben-Haim**
	Alex Friedlander
	Zaccheaus Oguntebi
	Patricia Yarbrough
Consultant for evaluation:	**Richard Shumway**
Consultants for development:	**Keith Hamann**
	John Wagner

Contents

Similarity and Equivalent Fractions

The Middle Grades Mathematics Project (MGMP) is a curriculum program developed at Michigan State University funded by the National Science Foundation to develop units of high quality mathematics instruction for grades 5 through 8. Each unit

- is based on a related collection of important mathematical ideas
- provides a carefully sequenced set of activities that leads to an understanding of the mathematical challenges
- helps the teacher foster a problem-solving atmosphere in the classroom
- uses concrete manipulatives where appropriate to help provide the transition from concrete to abstract thinking
- utilizes an instructional model that consists of three phases: launch, explore, and summarize
- provides a carefully developed instructional guide for the teacher
- requires two to three weeks of instructional time

The goal of the MGMP materials is to help students develop a deep, lasting understanding of the mathematical concepts and strategies studied. Rather than attempting to break the curriculum into small bits to be learned in isolation from each other, MGMP materials concentrate on a cluster of important ideas and the relationships that exist among these ideas. Where possible the ideas are embodied in concrete models to assist students in moving from the concrete stage to more abstract reasoning.

THE INSTRUCTIONAL MODEL: LAUNCH, EXPLORE, AND SUMMARIZE

Many of the activities in the MGMP are built around a specific mathematical challenge. The instructional model used in all five units focuses on helping students solve the mathematical challenge. The instruction is divided into three phases.

During the first phase the teacher *launches* the challenge. The launching consists of introducing new concepts, clarifying definitions, reviewing old concepts, and issuing the challenge.

The second phase of instruction is the class *exploration*. During exploration, students work individually or in small groups. Students may be gathering data, sharing ideas, looking for patterns, making conjectures, or developing other types of problem-solving strategies. It is inevitable that students will exhibit variation in progress. The teacher's role during exploration is to move about the classroom, observing individual performances and encouraging on-task behavior. The teacher urges students to persevere in seeking a solution to the challenge. The teacher does this by asking appropriate questions and by providing confirmation and redirection where needed. For the more able students, the teacher provides extra challenges related to the ideas being studied. The extent to which students require attention will vary, as will the nature of attention they need, but the teacher's continued presence and interest in what they are doing is critical.

When most of the students have gathered sufficient data, they return to a whole class mode (often beginning the next day) for the final phase of instruction, *summarizing*. Here the teacher has an opportunity to demonstrate ways to organize data so that patterns and related rules become more obvious. Discussing the strategies used by students helps the teacher to guide them in refining these strategies into efficient, effective problem-solving techniques.

The teacher plays a central role in this instructional model. The teacher provides and motivates the challenge and then joins the students in exploring the problem. The teacher asks appropriate questions, encouraging and redirecting where needed. Finally, through the summary, the teacher helps students to deepen their understanding of both the mathematical ideas involved in the challenge and the strategies used to solve it.

To aid the teacher in using the instructional model, a detailed instructional guide is provided for each activity. The preliminary pages contain a rationale; an overview of the main ideas; goals for the students; and a list of materials and worksheets. Then a script is provided to help the teacher teach each phase of the instructional model. Each page of the script is divided into three columns:

TEACHER ACTION	TEACHER TALK	EXPECTED RESPONSE
This column includes materials used, what to display on the overhead, when to explain a concept, when to ask a question, etc.	This column includes important questions and explanations that are needed to develop understandings and problem-solving skills, etc.	This column includes correct responses as well as frequent incorrect responses and suggestions for handling them.

Worksheet answers, when appropriate, and review problem answers are provided at the end of each unit; and for each unit test, an answer key and a blackline master answer sheet is included.

RATIONALE

From the many important topic areas in geometry, we have selected similarity and its close connections with equivalent fractions as the focus of the unit. We have chosen this topic because we consider it to be one of the more basic ideas in understanding the geometry of indirect measurement, proportional reasoning, scale drawing, scale models, and the nature of growth. Similarity provides an excellent opportunity to build more geometric meaning into the idea of equivalent fractions. When one constructs a scale model or drawing of a real object, the ratios of corresponding segments or distances for the scale model and the real object can be expressed as equivalent fractions.

The relationship between the *lengths* in similar figures and the *areas* of these similar figures is explored in the unit by looking at concrete examples and the patterns that come out of these examples. Looking at patterns and making generalizations from patterns is an important and appropriate problem-solving heuristic for students at this age. The culmination of the similarity unit is Activity 9, which poses for students a real application problem, the challenge of measuring the heights of inaccessible objects by using the mathematical concepts and models of the unit. All of these ideas are presented through concrete hands-on explorations for the students.

UNIT OVERVIEW

The unit begins with a very intuitive approach to enlargement, using the rubber band stretcher. While the scale and the shape are not precise, the basic idea stands out: If we produce an enlargement with two rubber bands knotted together, we get a figure whose sides are twice as long and whose area is four times as large as the original.

Enlargements are also the focus of the second activity. However, in this activity the use of a coordinate system allows more precise pictures. The scale factor between two similar figures is the factor by which the lengths are related. For example, if one figure has length two times as great as those of the smaller similar figure, the scale factor from the small to the large is two. The area of Morris's nose provides a concrete example of the relationship that area grows as the square of the scale factor. For example, if the sides are enlarged by a factor of three, the area is three squared, or nine times, as large. This is an important but subtle idea that will not be learned in this first exposure.

Activity 3 allows a close look at similarity of rectangles and develops a test that can be used to determine whether rectangles are similar. Students test two rectangles by comparing the short side of each rectangle to its long side. If the resulting ratios are equal, the rectangles are similar.

The idea of repeating tiles (reptiles) in Activity 4 brings together in one activity the ideas of similarity, congruence, and infinity. The student will realize that the reptiling process can go on infinitely in both directions. The tiles can continue to grow as large as one might want, and they can be continually subdivided as small as desired. Many students are fascinated by the ideas of infinity. Activity 4 defines similarity for any figures and lets students see, in a concrete way, the relationship of the growth of area to the growth of lengths.

In Activity 5, the students first apply what they learned in Activities 3 and 4 to similar right triangles (which are simply halves of the rectangles), and then to any similar triangles.

Activity 6 develops quick visual tests for similar rectangles and triangles. The students find that families of similar rectangles will have coinciding diagonals when the rectangles are nested in their lower left corner. Families of similar triangles will have corresponding sides parallel when nested at a corresponding angle. These tests are developed through an activity that emphasizes the physical manipulation of similar rectangles and triangles. The language of corresponding sides and angles, parallel lines, and coinciding lines (diagonals) is used.

Activity 7 provides a means of enlarging figures much more accurately than the stretcher in Activity 1. With an unmarked straight edge the students can construct point enlargements to various scale factors, both positive and negative. The effects of moving the point of enlargement determines where the image will lie but not how large it will be. The scale factor completely determines size.

Activity 8 provides a capstone to the work on area growth. The students pause and apply the ideas they have been studying. They can choose a picture to reproduce by any of the techniques they have learned so far. They can experiment to see which of the techniques provides the best results for their purposes. If you have access to a pantagraph (a mechanical linkage device for enlarging figures), you might introduce it at this point. The students measure lengths and areas to see how they grow. Finally they use their knowledge to solve some real application problems.

Activity 9 engages students in additional real-life applications of similarity to finding distances or heights that cannot be measured directly.

In the unit as a whole, students will meet the basic ideas and properties relating similar figures many times. It is not necessary that a student grasp each idea completely the first time it is met. The cyclic approach of the unit reinforces each idea as it is studied again in a slightly different setting.

Activity 1

THE VARIABLE TENSION PROPORTIONAL DIVIDER (STRETCHER)

OVERVIEW

The purpose of this activity is to introduce the idea of similar figures in an informal way. A simple device for drawing similar figures is made of two identical rubber bands knotted together. One end is anchored with the left hand and a pencil is inserted in the other end. The rubber band is stretched until the knot rests on the figure to be reproduced. The pencil is moved so that the knot traces the original figure. This will produce a similar figure whose area is four times as large as the original. Students discover by measuring that the lengths in the new figure are twice as long as the lengths in the original figure. Sometimes the new figure can be divided into four congruent figures that are all similar to the original. Students discover that area grows faster than length; if the new lengths are twice as long as the originals then the area is four times as large as the area of the original figure. These concepts will be developed further as the unit progresses.

Some students may be frustrated because the new drawings do not turn out perfectly. For this reason, much practice is provided before they are asked to look closely at the square, triangle, and rectangle in the summary.

More advanced students may wish to explore the effects of changing the anchor point (also called a *projection point*). They will find that changing the location of the point will change the location of the image but not the size of the image.

Goals for students

1. Draw similar figures using the variable tension proportional divider (stretcher).
2. Discover that a two-rubber-band stretcher gives a new figure that is similar to the original and whose corresponding lengths are twice as long.
3. Discover that a two-rubber-band stretcher gives a similar figure whose area is four times as large as the original.

Materials

Identical rubber bands; 2 per student (#18 rubber bands work well).
Blank paper.
Masking tape.
Rulers.

Worksheets

1-1, The Variable Tension Proportional Divider.
1-2, More Enlargements.
1-3, Summary Sheet.

THE VARIABLE TENSION PROPORTIONAL DIVIDER (STRETCHER)

TEACHER ACTION	TEACHER TALK	EXPECTED RESPONSE
On the board draw a picture of Pac Man with roughly a 10–15 cm diameter.	We are beginning a unit on similarity.	
	What do you think we mean when we use the word *similar*?	Two things are the same, or almost the same.
	Give me an example.	People are similar, cars, houses, clothes.
	Today we are going to learn about similarity using an ingenious little machine.	
Hold up two identical rubber bands. Knot two ends together.	It has two parts: We will knot the rubber bands like this.	
	This machine is called a *variable tension proportional divider*. We will just call it a *stretcher*.	
Point to Pac Man.	What is this strange figure?	Pac Man. Most students will recognize the figure from playing video games.
Demonstrate the machine.	Put one end of rubber band on the paper to the left of the figure far enough away so that the knot is on the figure. Call this point the *anchor point*.	
	Put pencil in the other end. Move pencil letting the knot trace the figure. Don't slack up on the first rubber band.	
	What happens?	We get another Pac Man similar to the original Pac Man, but bigger.
	What do we call this?	An enlargement.

Pencil

Knot

Anchor Point

Activity 1 *Launch*

TEACHER ACTION	TEACHER TALK	EXPECTED RESPONSE
Pass out rubber bands, blank sheets of paper, tape, and Worksheet 1-1, The Variable Tension Proportional Divider, and Worksheet 1-2, More Enlargements.	Make a stretcher by knotting the rubber bands together. Everyone should take two pieces of tape. Put these on the edge of your desk.	
Give directions on how to draw.	Tape Worksheet 1-1 down. Tape a blank sheet of paper next to it on the right.	
Have students use a blank sheet to copy figures 1 and 2. Then have them turn the sheet over for copying figure 3.	Keep the rubber band close to the end of the pencil *on the paper.* If you are left handed, turn the paper upside down and place the blank sheet on the left.	
Allow time for students to complete Worksheet 1-1.		

Worksheet 1-1   tape   blank sheet

tape

P ●

P ●

Activity 1 *Explore*

OBSERVATIONS

Encourage students to try a figure more than once to get a different enlargement.

For early finishers, suggest that Worksheet 1-2 be repeated with a three-rubber-band stretcher. This will require more paper. The students will have to be careful about placing the blank paper so that the finished figure will fit.

Another extension is to allow a chalkboard drawing to be enlarged. Students must choose an anchor point so that the enlargement stays on the board.

POSSIBLE RESPONSES

When the students finish these, have them try drawing the figures using a different anchor point and compare the results.

Students should discover that the location may change, but the size will remain the same.

8

Activity 1 *Summarize*

TEACHER ACTION	TEACHER TALK	EXPECTED RESPONSE
Pass out rubber bands and rulers and Worksheet 1-3, Summary Sheet.		
Ask. Allow students time to measure.	Very carefully, copy the square. How much larger do you think the new square is?	Various answers. It looks about twice as big.
	Let's measure the length of the original square and the new square and compare the lengths.	The measurements will be close to twice as large.
	How do the perimeters of the two squares compare?	The new perimeter is twice as large.
	If the larger square has a side twice as long, how many smaller squares will fit into it?	Many students will guess two.
Demonstrate how the larger square can be subdivided into four smaller squares that are congruent to the original.	1 1 Area = 1 2 2 Area = 4	
Ask similar questions for the triangle. The lengths of the new triangle are twice as large, and the area is four times as large. Demonstrate the subdivision.	Now copy the triangle and make the same measurements.	Area = 4

9

TEACHER ACTION	TEACHER TALK	EXPECTED RESPONSE
Explain.	The new figures that we created with the stretcher are similar to the original figures.	
	In mathematics, the word *similar* has a precise meaning. The activities in this unit will give us a chance to discover the mathematical meaning of similarity.	
Ask.	Copy the rectangle in figure 3. How does its perimeter and area compare with the original?	New perimeter = 16. New area = 12.
	How can you make a figure with sides 3 times as long?	Use three rubber bands.
	How will the new areas compare with the original?	New area = 9 × old area.
		Some students may need to see or actually make an enlargement of the square or triangle with three rubber bands to guess this relationship.

The Variable Tension Proportional Divider (Stretcher)

Use your stretcher to enlarge each figure. Use the anchor points that are marked.

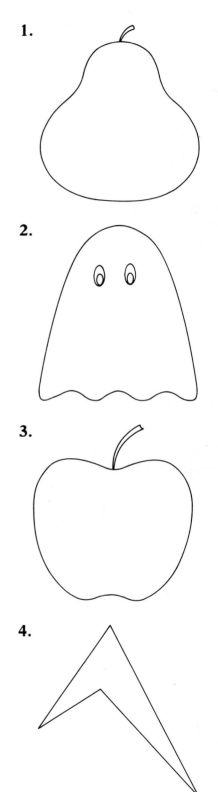

1.

2.

3.

4.

● **anchor point for 1 and 2**

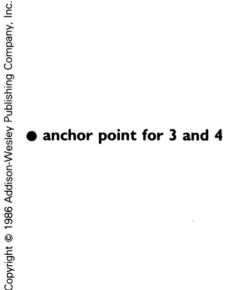

● **anchor point for 3 and 4**

More Enlargements

Use your stretcher to enlarge each figure. Use the anchor points (P) that are marked. Measure some corresponding parts of each figure you enlarge to see how the parts are growing.

1.

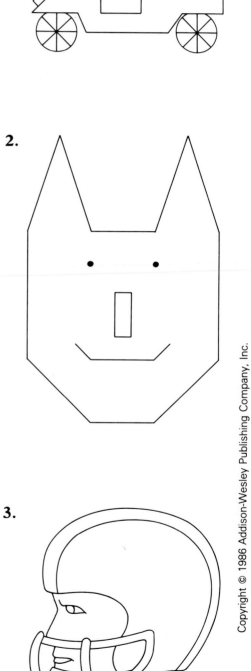

● **P for 1 and 2**

2.

3.

● **P for 3**

Worksheet 1-2

Summary Sheet

Use your stretcher to enlarge each figure.

1.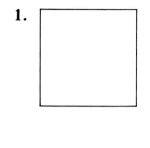

● **P for 1 and 2**

2.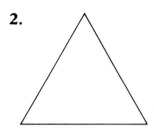

● **P for 3**

3.

Worksheet 1-3

Activity 2

MORRIS

OVERVIEW

In this activity students draw figures using a coordinate system. The first example produces a cat named Morris 1. Morris is transformed by stretching the coordinates into Morris 2, Morris 3, Boris, and Doris. Some are similar to Morris 1, and some are not. Similarity is intuitively defined as "being the same shape." Multiplying the coordinates of a figure by a constant and graphing the resulting points provides another scheme (in addition to the VTPD or Stretcher) for enlarging figures so that the "same shape" is preserved.

Students learn that in similar figures angle size is preserved, and all lengths are multiplied by a constant. Areas are multiplied by the square of that same constant. The fact that the ratios of corresponding sides in similar figures are equal is explored.

Goals for students

1. Transform figures by using a coordinate system.
2. Learn that angles and shape do not change in similar figures.
3. Learn how lengths and areas change in similar figures.
4. Meet the idea of classifying a rectangle by the ratio of its sides.

Materials

Straight edge.
*Drawings of Morris 1, 2, 3, Boris, and Doris (Materials 2-1).

Worksheets

*2-1, Morris
*2-1, page 4, Summary of Morris's Noses.
*2-2, Morris 99 (optional).

Transparencies

Starred items should be made into transparencies.

15

MORRIS

TEACHER ACTION	TEACHER TALK	EXPECTED RESPONSE
Put a 5 × 5 grid on the board.	Today we will start by playing a game of tic tac toe. How do we win tic tac toe?	Three in a row, etc.
	In this game we will win if we have four in a row, column, or diagonal.	
Use this point as 0,0 but do not label it.		
Divide the class into two teams, A and B.	Team A will have Xs; team B will get Os.	
The object is to have the class discover how to locate points on a grid by giving x and y coordinates.	To take a turn you get to say two numbers (and nothing else). Watch carefully to see how I use your numbers.	
Pick a student on team A.	Give me two numbers.	8, 10. (Numbers are given as examples.)
Starting at lower left corner, start to count to the right. Don't point out (0, 0); just always count from that point.	One, two, three, four,.... Oops! We fell off the chart.	
Select a member from team B.	Give me another pair of numbers.	2, 3.
Starting at the lower left corner, count *two* to the *right* and then *three up*. Mark this point with an O.	(Count as you move on the grid) One, two; one, two three.	

TEACHER ACTION	TEACHER TALK	EXPECTED RESPONSE
Repeat with a student from team A.	Give me another pair of numbers.	4, 3.
Count four to the right and then three up. Put an x at this point.		

Continue until one team has four symbols in a row, column, or diagonal or until there are no more moves. Be sure to give everyone a chance. If a team gives a point not on the grid or one that is already marked, it loses its turn.

Play two more games.

Students enjoy this game very much and may want to continue. The game is quickly played and can be used for the odd minute or two at the end of a period.

You may want to play the third game using the following grid. Start counting at center point (0, 0). Mark the x and y axis with a slightly darker chalk line, but don't explain. Let students figure out how to get to all places on the grid by watching you count out the numbers given at turns.

(0, 0)

TEACHER ACTION	TEACHER TALK	EXPECTED RESPONSE
Ask about the game 1 grid.	How did I use the pair of numbers to get a point?	For the first number you counted to the right and for the second number you counted up.
	Remember we will start at the lower left corner, which is called the *origin* or the point (0, 0).	
Pass out Worksheet 2-1 pages 1–3, Morris, and page 4, Summary of Morris's Noses.	Now we will use this grid to learn more about similar figures.	
Give directions. Plot the first two or three points together with the class.	We are going to find the points A–T on your grid paper.	
	How do we find A?	Start at 0. Go to the right five, and go up zero.
	How do we find B?	Right seven, up two.
	Draw a line from A to B.	
	Find C.	
	Connect these points with lines. Do this for points A–J. Start over for K–N, and for O–R and for S, and for T.	Many students get confused at first. Pacing them through several points will help.

Activity 2 *Launch*

TEACHER ACTION	TEACHER TALK	EXPECTED RESPONSE
When class finishes Morris 1 on Work-sheet 2-1, ask.	What does your figure look like?	A cat.
	What shape is Morris's nose?	Rectangle.
You may need to review what perimeter and area are. The grid picture is a good vehicle to review these concepts.	What is the perimeter of the rectangle?	6.
	What is the area of the rectangle?	2.
Give directions.	The points for Morris 2 are found from the Morris 1 points.	
	Point A on Morris 1 is (5, 0).	
	Point A on Morris 2 is $(2 \times 5, 2 \times 0)$ or (10, 0).	
	What is Point B on Morris 2?	(14, 4)
	What is Point B on Morris 3?	(21, 6)
	Draw Morris 2, Morris 3, Boris, and Doris.	
	Fill in the pairs of numbers before you draw a Morris.	

Activity 2 *Explore*

OBSERVATIONS

Students have to be careful to locate the correct points or the drawings will look wrong.

After drawing Morris 2, Morris 3, Boris, and Doris, students may start on the summary sheet.

Earlier finishers may do Morris 99, which is both an enlargement and a shift.

As an extra challenge, ask what happens to a picture if the order of the coordinates of each point is changed—plot each (x, y) as (y, x)?

POSSIBLE RESPONSES

Have students record coordinates of all the points for a Morris before plotting them.

When the order of the coordinates is changed, the picture will flip on its side.

Activity 2 *Summarize*

TEACHER ACTION	TEACHER TALK	EXPECTED RESPONSE
Ask.	How could we describe to a friend the growth of Morris 1, 2, and 3?	Various answers.
Put transparencies of Morris 1, 2, and 3 on the overhead without the grid (Materials 2-1, 2-2).		
Ask. Prod students to get as many things as possible.	As Morris 1 grew into 2 and 3, what things remained the same?	Straight lines remain straight lines; slant lines remain slant lines; horizontal and vertical lines remain horizontal and vertical; the nose remains centered between the eyes; ears are pointed; points remain points.
Students may need a hint to look at angle size.	What about the angles?	Some students will think the angles grow. Superimpose the ears on each other. Do the same for other angles.
Put a transparency of Boris and Doris on the overhead (Materials 2-1). Compare them with Morris 1, 2, and 3.	Do Boris and Doris have the same shape as Morris 1? How do they differ?	No. Boris is too fat; he grew in only one direction. Doris is too skinny; she also grew in only one direction.
	Are the angles in Boris and Doris the same as those in Morris 1?	No. The angles change.
	To have the "same shape," the corresponding angles must be the same size. Is keeping the corresponding angles equal enough to guarantee "the same shape"?	Various answers.
	Do all rectangles have the "same shape"?	No. Some are long and skinny, others are square. They have equal angles but are not all the same shape.
	This means that all rectangles are not similar to each other. We also need information on how the edges compare to determine when two figures are similar. It is not enough to just check angles. Angles are only one part of what makes figures have the same shape.	

Activity 2 *Summarize*

	TEACHER ACTION	TEACHER TALK	EXPECTED RESPONSE

TEACHER ACTION

Superimpose a Morris transparency one at a time on a transparency of a grid. This will help to find perimeter and area of the noses.

Show transparencies of Materials 2-1 (pictures of Morris) on the overhead as needed. Put a chart of distances and areas on the board (Worksheet 2-1, page 4, Summary of Morris's Noses).

Fill in all columns for Morris 1, 2, and 3.

Solicit the entries for all columns. Have students fill in their copies.

For younger students, perimeter patterns could be postponed until after the discussion of the area growth. This makes it a bit easier to see the pattern between the multiplier (or scale factor) and the growth of the area.

TEACHER TALK

We discussed what things did not change. Let's see what changes as Morris 1 grows.

Morris	Rule	Bottom Edge	Side Edge	Ratio: $\dfrac{\text{Bottom}}{\text{Side}}$	Area of nose	Perimeter of nose
1	(x, y)	1	2	$\dfrac{1}{2}$	2	6
2	$(2x, 2y)$	2	4	$\dfrac{2}{4} = \dfrac{1}{2}$	8	12
3	$(3x, 3y)$	3	6	$\dfrac{3}{6} = \dfrac{1}{2}$	18	18
4	$(4x, 4y)$	4	8	$\dfrac{4}{8} = \dfrac{1}{2}$	32	24
5	$(5x, 5y)$	5	10	$\dfrac{5}{10} = \dfrac{1}{2}$	50	30

Tell me two Morrises that are similar.

Look at our record sheet. Which columns are measures of lengths?

What happened to the lengths when we went from Morris 1 to Morris 2?

What happened to the lengths when we went from Morris 1 to Morris 3?

What happened when we formed a ratio of the bottom edge over the side edge?

Now we can give "same shape" or *similar* a better definition. Two figures are similar if their corresponding angles are equal *and* ratios of corresponding sides are equal.

EXPECTED RESPONSE

Morris 1 and Morris 2
Morris 1 and Morris 3
Morris 2 and Morris 3

Edges of nose, perimeter of nose.

They doubled.

Three times as much.

All the ratios were equal.

Activity 2 *Summarize*

TEACHER ACTION	TEACHER TALK	EXPECTED RESPONSE
Ask.	What about the areas of the nose? Going from Morris 1 to Morris 2 what happened?	Times 4. If students say "plus 6," ask for the change in terms of *times* something.
	Going from Morris 1 to Morris 3?	Times 9.
	Suppose we draw a Morris 4 using the rule $(4x, 4y)$.	
	What will the numbers be for the columns in the summary sheet?	$(4x, 4y)$ 4 8 $\frac{1}{2}$ 32 24
		If students have difficulty with area, draw the nose:
Fill in the row for Morris 4.		
Fill in the row for Morris 5.	If we draw a Morris 5 using the rule $(5x, 5y)$, what will the numbers be for the summary sheet?	$(5x, 5y)$ 5 10 $\frac{1}{2}$ 50 30
	Will the rule $(3x, 4y)$ produce a cat similar to Morris 1?	No; you must multiply both coordinates by the same factor.
Refer to the chart.	Suppose that in the column marked perimeter there is the number 42. What is the rule for producing a similar Morris whose nose has perimeter 42.	$(7x, 7y)$
	If the bottom edge is 10, what rule produces a similar Morris?	$(10x, 10y)$
Make list. Then ask about $(8x, 8y)$ and $(10x, 10y)$.	Look at nose area for similar Morrises. Tell me the area if the rule is (x, y).	2
	What if the rule is $(2x, 2y)$?	$8 = 4 \times 2$
	What if the rule is $(3x, 3y)$?	$18 = 9 \times 2$
	What if the rule is $(4x, 4y)$?	$32 = 16 \times 2$
	What if the rule is $(8x, 8y)$?	$128 = 64 \times 2$
	What if the rule is $(10x, 10y)$?	$200 = 100 \times 2$

23

Activity 2 *Summarize*

TEACHER ACTION	TEACHER TALK	EXPECTED RESPONSE
It is important to stress that in similar figures all the linear measurements in a figure grow by the same *multiples*; areas grow by *square numbers*.	In each case, if we multiplied the bottom edge and side edge by a number, we got a new figure, similar to the original, with an area that is the number squared times as large. If we multiplied by 9, the new figure has an area 9^2, or 81 times as large as the original.	
(Optional)	What rules would shrink Morris 1? What if the rule is $(\frac{1}{2}x, \frac{1}{2}y)$? What are the numbers?	Use fractions or use a smaller grid. $\left(\frac{1}{2}x, \frac{1}{2}y\right)$ $\frac{1}{2}$ 1 $\frac{1}{2}$ 3
	This is an example of an ''enlargement'' that *shrinks* the original figure.	These numbers may be hard for the students to see—especially the area, which is $\frac{1}{2} \times \frac{1}{2} = \frac{1}{4}$ times as large as the original Morris 1.
Challenge some students to make up a figure of their own with the appropriate rule for growing it. Students could exchange rules.		
Students could use a grid to enlarge or shrink their favorite cartoon character.		
Use Worksheet 2-2, Morris 99, as a challenge.		
You may want to have students transfer a drawing to a grid and then enlarge it by using a $(4x, 4y)$ transformation.		
You may want to have students transfer a drawing to a square grid and draw it again, using the same coordinates on a grid with larger squares.		

Morris

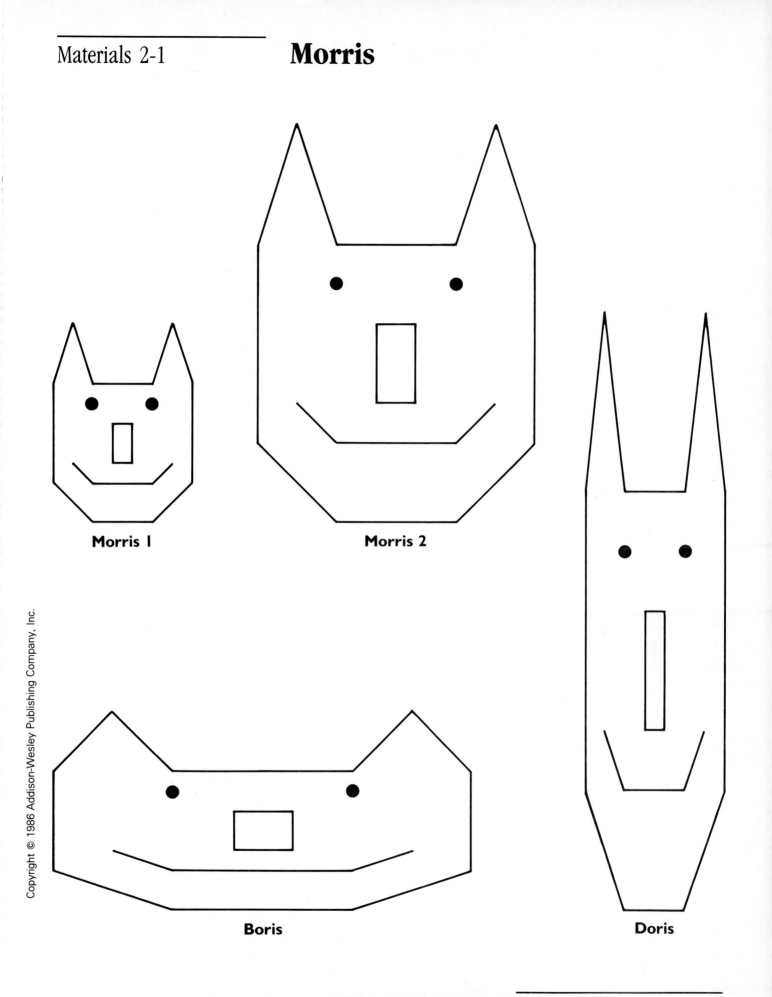

Morris 1

Morris 2

Boris

Doris

Morris

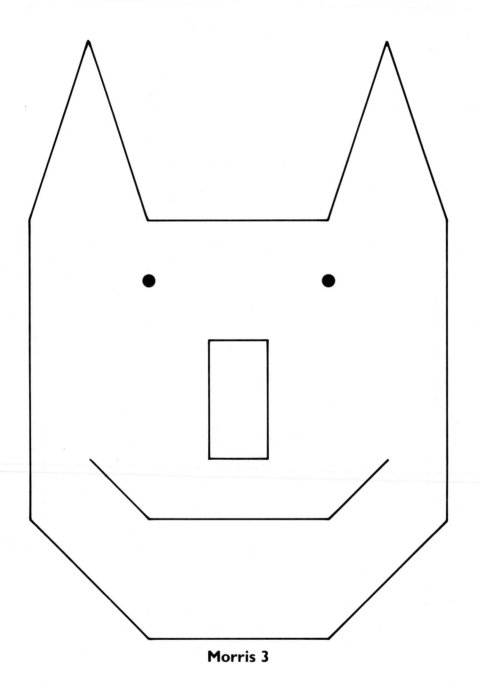

Morris 3

Morris

1. From the column marked Morris 1, plot the points on the grid paper. Connect the points in the first set with lines. Do the same for sets II & III.

2. When you finish Morris 1, do the others on the corresponding grid paper. Write in all the pairs of numbers in the column first.

3. For the others, go back to Morris 1 to get the points.

Points		Morris 1	Morris 2	Morris 3	Boris	Doris
		(x, y)	$(2x, 2y)$	$(3x, 3y)$	$(3x, y)$	$(x, 3y)$
Set I	A	5, 0				
	B	7, 2				
	C	7, 7				
	D	6, 10				
	E	5, 7				
	F	2, 7				
	G	1, 10				
	H	0, 7				
	I	0, 2				
	J	2, 0				
		Connect to A				
		Start Over				
Set II	K	1, 3				
	L	2, 2				
	M	5, 2				
	N	6, 3				
		Start Over				
Set III	O	3, 3				
	P	4, 3				
	Q	4, 5				
	R	3, 5				
		Connect to O				
Set IV	S	5, 6 (DOT)				
Set V	T	2, 6 (DOT)				

Morris

Morris I

Morris 2

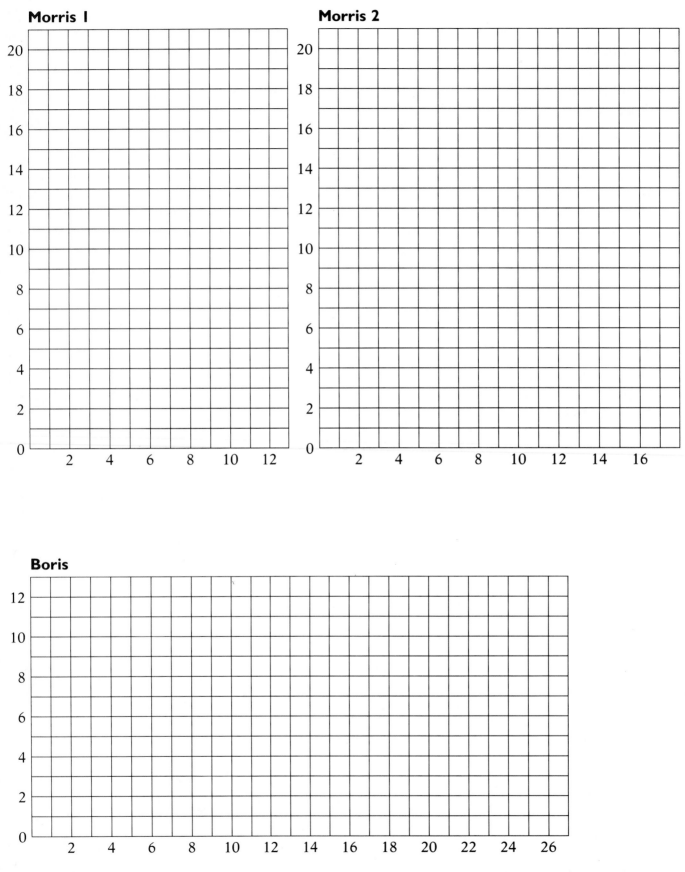

Boris

Worksheet 2-1, page 2

Morris

Morris 3

Doris

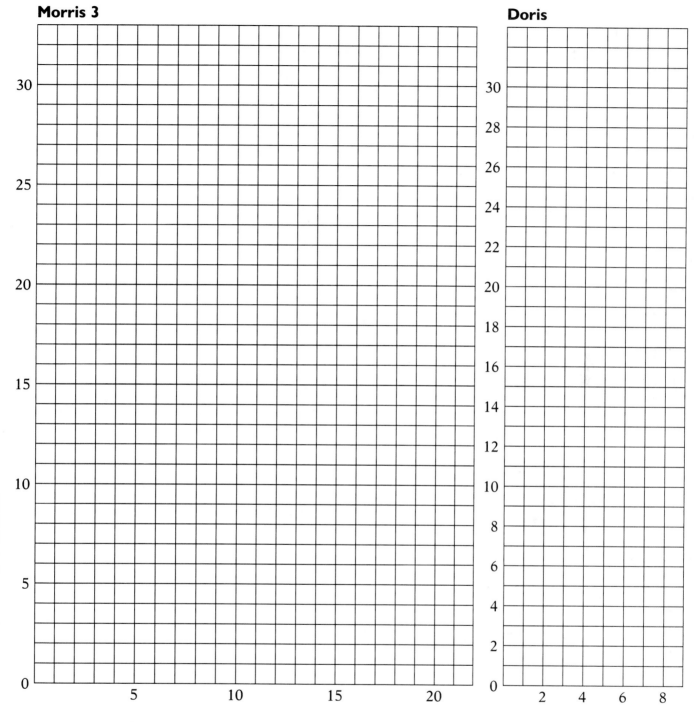

Summary of Morris's Noses

Morris	Rule	Bottom Edge	Side Edge	Ratio: Bottom/Side	Area	Perimeter
1	(x, y)					
2	$(2x, 2y)$					
3						
4						
5						
						42
		10				
Boris						
Doris						

Do Boris and Doris fit the same patterns as Morris 1, 2, and 3? _____

Does Morris 99 fit in the same patterns as Morris 1, 2, and 3? _____

Fill in the missing spaces on the rows with perimeter 42 and bottom edge 10.

Worksheet 2-1, page 4

Morris 99

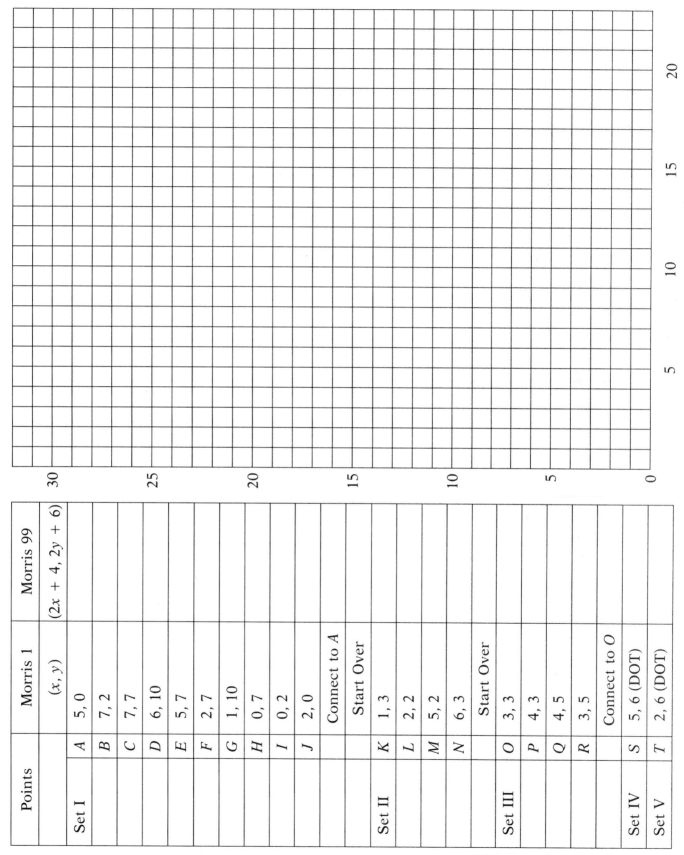

Points		Morris 1	Morris 99
		(x, y)	$(2x + 4, 2y + 6)$
Set I	A	5, 0	
	B	7, 2	
	C	7, 7	
	D	6, 10	
	E	5, 7	
	F	2, 7	
	G	1, 10	
	H	0, 7	
	I	0, 2	
	J	2, 0	
		Connect to A	
		Start Over	
Set II	K	1, 3	
	L	2, 2	
	M	5, 2	
	N	6, 3	
		Start Over	
Set III	O	3, 3	
	P	4, 3	
	Q	4, 5	
	R	3, 5	
		Connect to O	
Set IV	S	5, 6 (DOT)	
Set V	T	2, 6 (DOT)	

SIMILAR RECTANGLES

OVERVIEW

In this activity, rectangles are classified into families by the ratio $\frac{\text{short side}}{\text{long side}}$. For similar rectangles, the ratios are equivalent fractions.

Students are given several rectangles and asked to find which are similar by measuring the sides and computing the ratios of short side to long side. Transparent grid paper is used to help emphasize the meaning of area and perimeter when the students are measuring. In the summary students find missing lengths of sides using the fact that two rectangles are similar.

Goals for students

1. Learn that in similar rectangles the ratios $\frac{\text{short side}}{\text{long side}}$ are equal.
2. Measure rectangles and decide which are similar.
3. Review area and perimeter of rectangles.

Materials

*Transparent grid paper (Materials 3-1).
*Morris's Noses (Materials 3-2).

Worksheets

*3-1, Which Rectangles Are Similar?
*3-2, Similar Rectangles.
 3-3, Ratios and Similar Rectangles.

Transparencies

Starred items should be made into transparencies.

MATHEMATICS DEPARTMENT
ALVERNO COLLEGE
MILWAUKEE, WI 53234-3922

TEACHER ACTION	TEACHER TALK	EXPECTED RESPONSE
Put a transparency of Morris's Noses (Materials 3-2) on the overhead.	Let's look at Morris's nose. Which of the noses (rectangles) have the same shape as Morris 1?	2, 3
	What do we call these figures that have the same shape?	They are similar.
Some students may need to review equivalent fractions and ratio before going on with Activity 3.	What things are equal in similar rectangles?	Angles and ratios of corresponding sides.
	What things can change in similar rectangles?	Lengths and area.
	Lengths and area change in similar figures, but in a very special way.	
	Let's look at the ratio of the short side to the long side. The short side is a and the long side is b.	Ratio: $\dfrac{a}{b} = \dfrac{\text{short side}}{\text{long side}}$
	What are the ratios in the rectangles for Morris's noses from Activity 2?	Morris 1: $\dfrac{1}{2}$
		Morris 2: $\dfrac{2}{4} = \dfrac{1}{2}$
		Morris 3: $\dfrac{3}{6} = \dfrac{1}{2}$
		Boris: $\dfrac{2}{3}$
		Doris: $\dfrac{1}{6}$

3

2

Boris

1

Doris

TEACHER ACTION	TEACHER TALK	EXPECTED RESPONSE
	In similar rectangles the ratios $\frac{short\ side}{long\ side}$ are equivalent fractions.	
	Which of these noses are similar?	1, 2, and 3.
	We can also use this as a test for similar rectangles. If the ratios are equal, then the rectangles are similar.	
	We said earlier that similar figures must also have angle measure preserved. Why don't we need to check angles on rectangles?	All the angles of rectangles are 90°.
Students need Worksheet 3-1, Which Rectangles Are Similar? and Worksheet 3-2, Similar Rectangles, and grid paper.	All of these figures are rectangles. Your task is to find which ones are similar.	
Put a transparency of Worksheet 3-1, Which Rectangles Are Similar? on the overhead.		
Ask.	How could we do this?	Look for one with the same shape. (If this response is given, say it is not always easy to use "same shape" as a test. Rectangles all look much alike. Ask how to be sure the shape is the same.)
	How else could we do it?	Find the ratios of short side to long side. If the ratios are equal, then the rectangles are similar.
	Could we use the ratio of long side to short side?	Yes, just as long as we are consistent.

35

Activity 3 *Launch*

TEACHER ACTION	TEACHER TALK	EXPECTED RESPONSE
Use a one-fourth page transparency of grid paper (Materials 3-1).	In this activity you will collect some other information about each rectangle, its perimeter and its area.	
Demonstrate how to use the grid as a ruler to find the lengths and areas. Place the grid over the figure and count squares for area and units for length.	What is the perimeter of a rectangle?	The distance around.
	How do we find the perimeter using the grid?	Count the lengths on the transparency.
	How do we find the area?	Count all the squares in the rectangle or multiply the two sides together: $A = W \times L$ or $A = $ long side $\times$ short side.

Fill in the first row of the chart on
Worksheet 3-2.

		Short Side *a*	Long Side *b*	Ratio $\frac{a}{b}$	Area	Perimeter
Rectangles	1	3 units	4 units	$\frac{3}{4}$	12 Square units	14 units

Give students time to collect data and fill
in the chart.

Activity 3 *Explore*

OBSERVATIONS	POSSIBLE RESPONSES
Some students will make the mistake of trying to count a row of squares when measuring perimeter.	Use the language of walking around the block or wrapping a string around the rectangle.
Some students confuse area and perimeter.	Stress the basic difference in units of measurement for length and area. Be sure groups of students are focusing on "distance around" and "covering" as they measure the rectangles. Ask a question such as, "George measured a rectangle and got 12. What could that . . . mean?"

Activity 3 *Summarize*

	TEACHER ACTION	TEACHER TALK	EXPECTED RESPONSE

TEACHER ACTION

Put a transparency of Worksheet 3-2, Similar Rectangles, on the overhead and collect the data from the class for rectangles.

Ask.

TEACHER TALK

Rectangles	Short side a	Long side b	Ratio $\frac{a}{b}$	Area	Perimeter
1	3 units	4 units	$\frac{3}{4}$	12 sq. units	14 units
2	6	8	$\frac{6}{8} = \frac{3}{4}$	48	28
3	9	12	$\frac{9}{12} = \frac{3}{4}$	108	42
4	12	16	$\frac{12}{16} = \frac{3}{4}$	192	56
5	15	20	$\frac{15}{20} = \frac{3}{4}$	300	70
6	10	14	$\frac{10}{14} = \frac{5}{7}$	140	48

Which rectangles are similar?

Could we have decided this by looking at just the shape? Why?

How did you decide which rectangles are similar?

We found the ratio $\frac{a}{b}$ or the short side to the long side. Could we use $\frac{b}{a}$?

What happens?

Why do we not have to check angles?

EXPECTED RESPONSE

Rectangles 1, 2, 3, 4, and 5.

No; same shape is ambiguous, hard to tell.

Check the ratio of the sides: $\frac{\text{short side}}{\text{long side}}$.

Yes we could use $\frac{b}{a}$ provided we use it for all rectangles. $\frac{b}{a}$ is an improper fraction; $\frac{b}{a}$ is larger than 1.

$$\frac{a}{b} = \frac{3}{4}$$

$$\frac{b}{a} = \frac{4}{3}$$

Because rectangles have all angles equal to 90°.

Activity 3 *Summarize*

TEACHER ACTION	TEACHER TALK	EXPECTED RESPONSE
Ask.	Let's put some more similar rectangles on the list.	
Enter the new rectangles.	If the short side is 30, what's the long side of the similar rectangle?	40; $\frac{30}{40} = \frac{3}{4}$.
	If the long side is 32, what's the short side?	24
	What are some other rectangles that are similar to the rectangles with ratio of sides $\frac{3}{4}$?	Various answers.
	Let's look at another pair of similar rectangles.	
Put the following rectangles on the board or overhead.	What is the missing side?	10
	What do you multiply the smaller rectangle's edges by to get the larger rectangle? This is called the *scale factor* from the smaller to the larger rectangle. It tells us what to multiply the dimensions of the original by to get the new figure.	5
	What is the missing side?	4
Show these.	What's the perimeter of the larger rectangle?	24
	What's the perimeter of the smaller rectangle?	12
	What is the *scale factor* from the larger to the smaller rectangle? This means that we multiply the larger (the original) dimensions by $\frac{1}{2}$ to get the dimensions of the new figure.	$\frac{1}{2}$

Rectangles (Put the following rectangles on the board or overhead):

- small rectangle: 2 by 3
- large rectangle: 15 by ?

Rectangles (Show these):

- rectangle: ? by 8
- rectangle: 2 by 4

TEACHER ACTION	TEACHER TALK	EXPECTED RESPONSE
	What about these? If the figures are similar, what is the missing side?	6 units. If students have difficulty, write the equivalent fractions $$\frac{4}{6} = \frac{?}{9}$$ and suggest they rewrite $\frac{4}{6}$ in lowest terms.
If students are familiar with finding equal fractions, this part should go quickly. Otherwise this is a good time to review or reinforce equivalent fractions: renaming to lower terms or to higher terms.	When the figures are given to be similar, what can we assume about corresponding angles?	They are equal.
	About the ratios of corresponding sides?	They are equal.
Distribute Worksheet 3-3, Ratios and Similar Rectangles. (You may prefer to assign this as homework.)	Here are some practice problems that use the ideas about similarity that we have studied.	

$\frac{1}{2}$-cm Grid

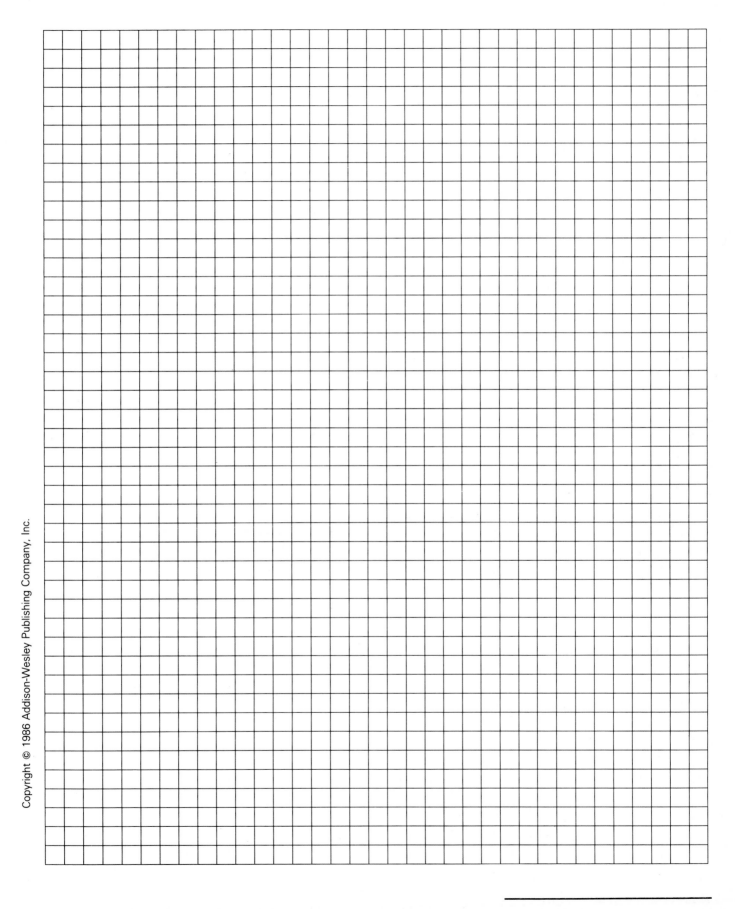

Morris's Noses

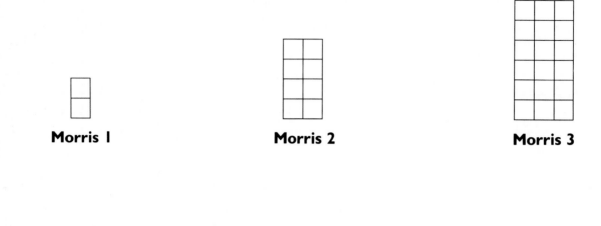

Morris 1   **Morris 2**   **Morris 3**

Boris   **Doris**

Which Rectangles Are Similar?

Use your grid to measure each rectangle. Record the information on Worksheet 3-2, Similar Rectangles. Determine which of the rectangles are similar.

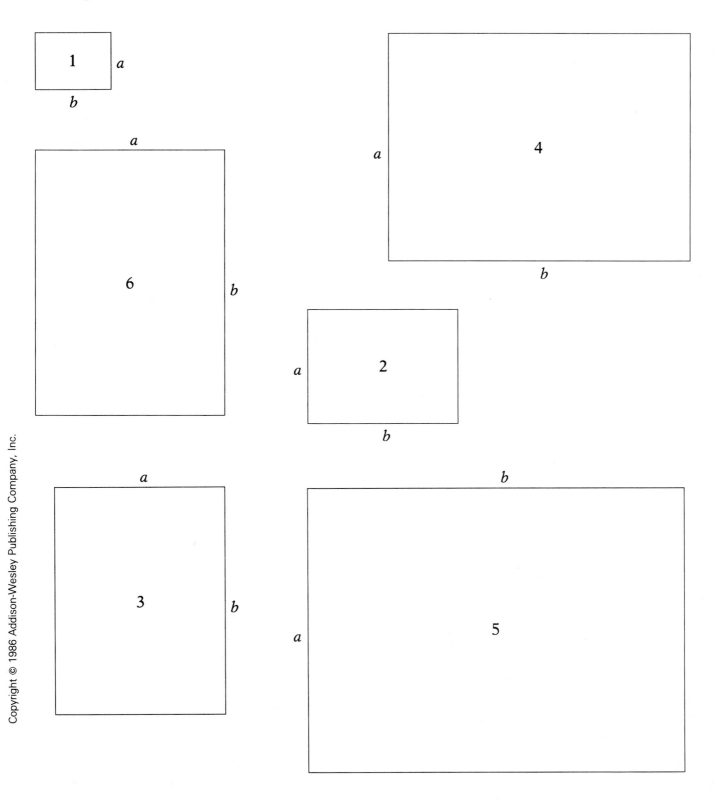

Similar Rectangles

Rectangle	Short side a	Long side b	Ratio $\frac{a}{b}$	Area	Perimeter
1					
2					
3					
4					
5					
6					

Which rectangles are similar? _____

Give a rule for testing rectangles to see if they are similar. _____

_____

_____

Worksheet 3-2

Ratios and Similar Rectangles

Without measuring, which pairs of rectangles (1–4) are similar? _____

1. 2 | 3 | 6 | 9

2. 6 | 2 | 3 | 2

3. 8 | 10 | 4 | 5

4. 8 | 4 | 10 | 6

Each pair of rectangles (5–10) are similar. Find the missing measurements.

5. Side a = _____   **6.** Side a = _____

3 | 5 | 20 | a 1 | 4 | 8 | a

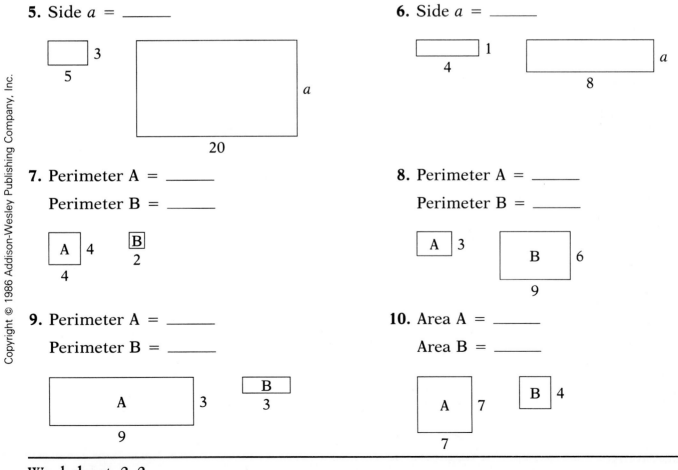

7. Perimeter A = _____   **8.** Perimeter A = _____

Perimeter B = _____   Perimeter B = _____

A | 4 | 4 | B | 2   A | 3 | B | 6 | 9

9. Perimeter A = _____   **10.** Area A = _____

Perimeter B = _____   Area B = _____

A | 3 | 9 | B | 3   A | 7 | 7 | B | 4

Activity 4

REPTILES

Reptiles (short for repeating tiles) are shapes (tiles) that can be subdivided according to certain rules. The two rules we use in this unit are

1. All subdivided parts must be *congruent* to each other (having exactly the same shape *and* same size as each other).
2. All subdivided parts must be *similar* to the original shape (having exactly the same shape as the original).

Finding reptiles is an interesting way of examining the relationship of area between similar figures. This special type of tiling is a reversible process: a figure can be subdivided (reptiled) into congruent figures each similar to the original figure. Conversely, congruent shapes can be put together to form a figure that is similar to the smaller figures. If we begin with a reptile, the figure can always be built up or subdivided into $4, 16, 64, \ldots 4^n$ figures. Often the figures can also be subdivided into 9, 25, 36, 49, 81, etc., congruent figures; this reinforces the idea that the area grows as the square of the factor by which the sides grow.

This activity provides indirect experiences with testing angles of similar figures to see that the measure of angles is preserved. Students are exposed to many drawings of figures of triangles subdivided into reptiles. These drawings have in them many examples of parallel lines and superimposed angles that lay the groundwork for understanding the diagonal test for similar triangles that is introduced in Activity 5.

Goals for students

1. Distinguish between congruent and similar shapes.
2. Put four congruent shapes together to form a larger figure that is similar to each of the four smaller figures (original figure).
3. Subdivide a figure into four congruent parts, each similar to the larger (original) figure.
4. Recognize that a figure can be subdivided into congruent shapes each similar to the original figure. The number of subdivisions must be a square (i.e., 4, 9, 16, 25, etc.).
5. Recognize that congruent shapes can be put together to form a larger similar figure.
6. Determine that if a figure is enlarged by a scale factor, s, the sides of the new figure are s times as long as the original and the area is s^2 times as big.

Materials

Plastic (or paper) geometric shapes (*Geometric Figures, Materials 4-1, two sheets per student).

Worksheets

*4-1, Reptiles I.
*4-2, Reptiles II.
 4-3, Super Reptiles.
 4-4, Are They Similar?

Transparencies

Starred items should be made into transparencies.

TEACHER ACTION	TEACHER TALK	EXPECTED RESPONSE
Put a set of plastic or paper geometric figures on the overhead.		
Ask.	What are the names of these shapes?	Parallelograms, triangles, rectangles, trapezoids, and hexagons.
Separate out three or four congruent triangles.	What is true about these figures? How do their size and shape compare?	They are all the same. They fit exactly on top of each other.
	What do we call these figures that have the same *shape* and same size?	*Congruent* triangles.
Put on two similar non-congruent pieces	Are these congruent? Why?	No; shape is the same, but the size is different.
Demonstrate other congruent, non-congruent, similar, and nonsimilar shapes.	Which of these are congruent? Which of these are similar?	All congruent shapes are similar but not vice versa.
Pass out a bag of geometric shapes or pass out two sheets of Geometric Figures (Materials 4-1); have students cut out figures first.	Arrange these in congruent groups.	
Allow time for students to become familiar with the materials.		
Put four rectangles together on the overhead. Demonstrate.	Do you think I could put these four rectangles together to make another rectangle that is similar to these smaller rectangles?	Yes.
	We call this a repeating tile, or a reptile.	
	Use four of each shape. Try to put them together to make a larger figure that is similar to the smaller figures.	

OBSERVATIONS

POSSIBLE RESPONSES

Some will tile hexagons like this:

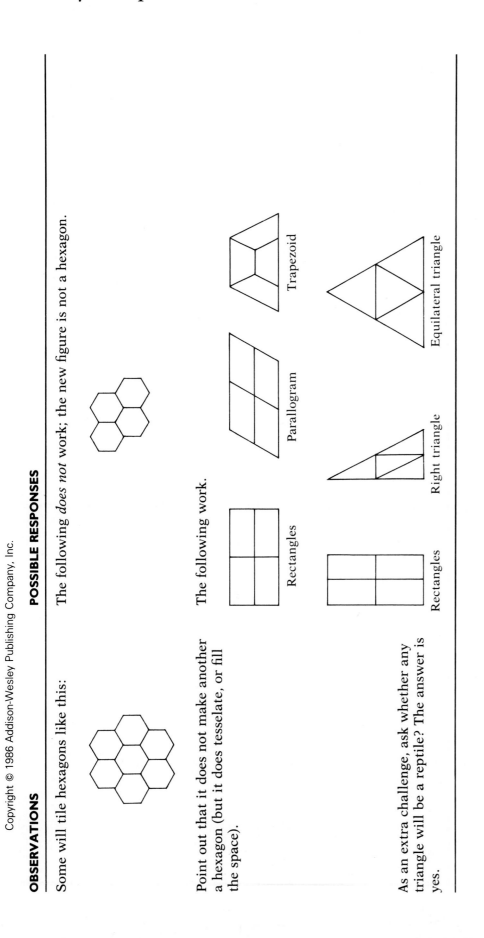

Point out that it does not make another a hexagon (but it does tesselate, or fill the space).

As an extra challenge, ask whether any triangle will be a reptile? The answer is yes.

The following *does not* work; the new figure is not a hexagon.

The following work.

Rectangles

Parallelogram

Trapezoid

Rectangles

Right triangle

Equilateral triangle

Activity 4 *Summarize*

TEACHER ACTION	TEACHER TALK	EXPECTED RESPONSE
Have a student demonstrate a reptile on the overhead.	Which of the figures work?	All triangles, squares, rectangles, parallelograms always work.
Display.	Rectangles work this way.	
	Do they work this way?	Only if $\frac{\text{short side}}{\text{long side}} = \frac{1}{2}$
Display.	Parallelograms work this way and also	
	this way if $\frac{\text{short side}}{\text{long side}} = \frac{1}{2}$.	

TEACHER ACTION	TEACHER TALK	EXPECTED RESPONSE
Display.		
Ask.	How did we define similar figures in the Morris activity?	Similar figures have equal corresponding angles and equal ratios of corresponding sides.
	How do we show that the big trapezoid is similar to each of the small ones?	Check the angles and measure the sides.
Demonstrate that the angles are the same by superimposing a copy of the small trapezoid on each angle of the big one.	Let's put a copy of the trapezoid on top of the big one to see if the angles stay the same.	
	Yes, they do.	
	Now what about the sides?	The long base of the trapezoid is twice one of the slant sides and twice the short base. So the new base is twice the old base and the new sides are twice the old sides. The ratio is $\frac{1}{2}$ in each case.
	Good. That means that the big trapezoid is similar to the original small ones and is therefore a reptile.	
If necessary, review that *isosceles* means that the two slant sides of the trapezoid are equal.	Notice that isosceles trapezoids work only when the top and the sides are the same length and the bottom is twice that length.	

Activity 4 *Summarize*

TEACHER ACTION	TEACHER TALK	EXPECTED RESPONSE
Check a few results. Any four congruent triangles can be put together to form a larger, similar triangle.	Draw a triangle different from the ones you were given. Cut it out and cut three more congruent to it. Put these four triangles together to make a larger, similar triangle.	Students can cut all four triangles by folding a sheet into quarters and cut it all at once.
Put four congruent triangles on the overhead. Ask a student to come up and put them together to form a similar triangle. Pull the four triangles slightly apart.	How do the lengths of the bases compare? If the sides of the larger figure are twice the length of the smaller figure, then we say the smaller figure grows by a *scale factor* of 2.	The base of the larger triangle is twice as long as the base of the smaller triangle.
	How do the areas compare?	The area of the larger is four times the smaller.
Add another row to the large triangle on the overhead.	When we add another row of similar triangles to the figure, how do the base and area of the large triangle compare to one of the small triangles?	The base is 3 times as large, the area is 9 times as large.
	What is the scale factor and why?	3, because the base grows by a factor of 3.
	If I want to add another row of triangles and still have a similar triangle, how many small triangles will it take?	There may be several guesses. Try some until you get the correct answer of 7.
	How does the base and area of the large triangle compare to the base and area of one of the smaller triangles?	The base is 4 times as large. The area is 16 times as large.
	What is the scale factor and why?	4, because the base grows by a factor of 4.

Base

Base

Scale Factor 3

Activity 4 *Summarize*

TEACHER ACTION / TEACHER TALK	EXPECTED RESPONSE
Let's make a chart to show what we have found.	Scale Factor — Area in Triangles 2 — 4 3 — 9 4 — 16
What is the relationship between the scale factor and how the area grows?	The area growth is the scale factor squared.
If the scale factor is 5, how many triangles would it take to build the new figure?	$5^2 = 25$ triangles.
If the number of triangles to cover the area is 144, what would the scale factor be?	$12^2 = 144$; scale factor would be 12.
What would a scale factor of 1 mean?	The figure is the same.
If we organize the data we gathered in another way, we can see another interesting pattern. Let's show the number of triangles needed to grow each new large triangle.	New Triangles Added — Total Triangles 1 — 1 3 — 4 5 — 9 7 — 16 . . . — . . .
How many triangles would the next row take?	9 new triangles
And how many triangles would we have all together?	25 new triangles
Do you see a pattern that would help us add up odd numbers?	The sum of the first N odd numbers is N^2. Example: $1 + 3 + 5 + 7 + \ldots + 37 = 19^2$ 19 odd numbers
You can leave this as an extra challenge pattern or help the class see that the sum of the first N odd numbers is N^2.	

Scale Factor 4

53

Activity 4 *Summarize*

TEACHER ACTION	TEACHER TALK	EXPECTED RESPONSE
	We have built reptiles by adding on figures. Now let's go the other way.	
Draw a triangle on the overhead.	Who can show how to reptile this triangle by subdividing it? How did you get your answer?	Various answers. Eventually try to elicit "For any triangle, bisect each side and connect the three points."
Subdivided.		
Ask.	We can subdivide each of these four triangles into four smaller congruent triangles that are similar to the original triangle. Can we subdivide it into more than four smaller congruent triangles that are similar to the larger? How?	Yes. Divide each of the four triangles into four more to get 16.
Pass out Worksheet 4-1, Reptiles I.	Here are some figures to be subdivided into *four* congruent parts. Remember, each must be similar to the original figure.	

Activity 4 *Explore*

OBSERVATIONS

POSSIBLE RESPONSES

Many students will try to subdivide number 5 into squares.

Remind them the small pieces must be *similar* to the original. One solution is

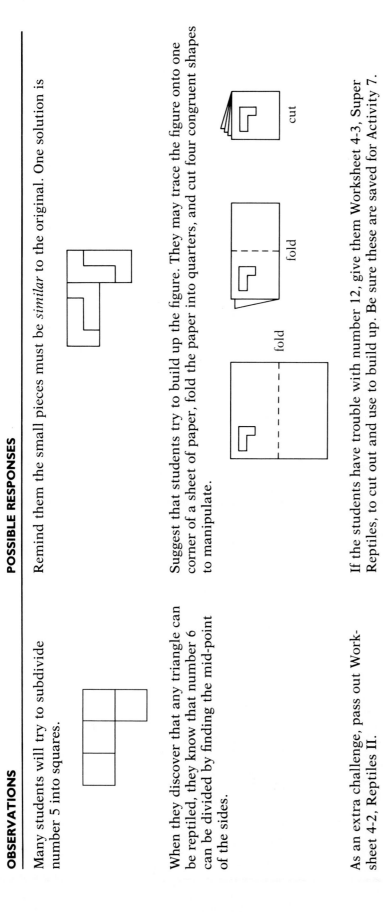

When they discover that any triangle can be reptiled, they know that number 6 can be divided by finding the mid-point of the sides.

Suggest that students try to build up the figure. They may trace the figure onto one corner of a sheet of paper, fold the paper into quarters, and cut four congruent shapes to manipulate.

As an extra challenge, pass out Work-sheet 4-2, Reptiles II.

If the students have trouble with number 12, give them Worksheet 4-3, Super Reptiles, to cut out and use to build up. Be sure these are saved for Activity 7.

Activity 4 *Summarize*

TEACHER ACTION	TEACHER TALK	EXPECTED RESPONSE
Ask.	Who can display a solution to number 1?	
Continue until all solutions are displayed.		
As an extra challenge, have students subdivide a triangle into 9 congruent parts, another into 16 congruent parts.		
Ask the extra challenge question on Worksheet 4-3, Super Reptiles.	How many of the figures marked 1 are needed to cover the similar figure marked 3?	The answer is 9.
Have students give examples to illustrate or test their statements.	What have we learned about similar figures?	Ratios of corresponding sides are equal; corresponding angles are the same size; if the scale factor is S, the area is S^2 times as large.
Pass out Worksheet 4-4, Are They Similar? This could be assigned as homework.	For each pair of figures on Worksheet 4-4, say whether or not they are similar and why or why not.	
Students can either measure angles or trace and compare. The figures are drawn so that measuring with the half centimeter grids works well.	Remember that you must check to see that ratios of corresponding sides are equal and that corresponding angles are equal.	

Geometric Figures

You should have two sheets like this one. Cut apart the figures so that you have four of each shape.

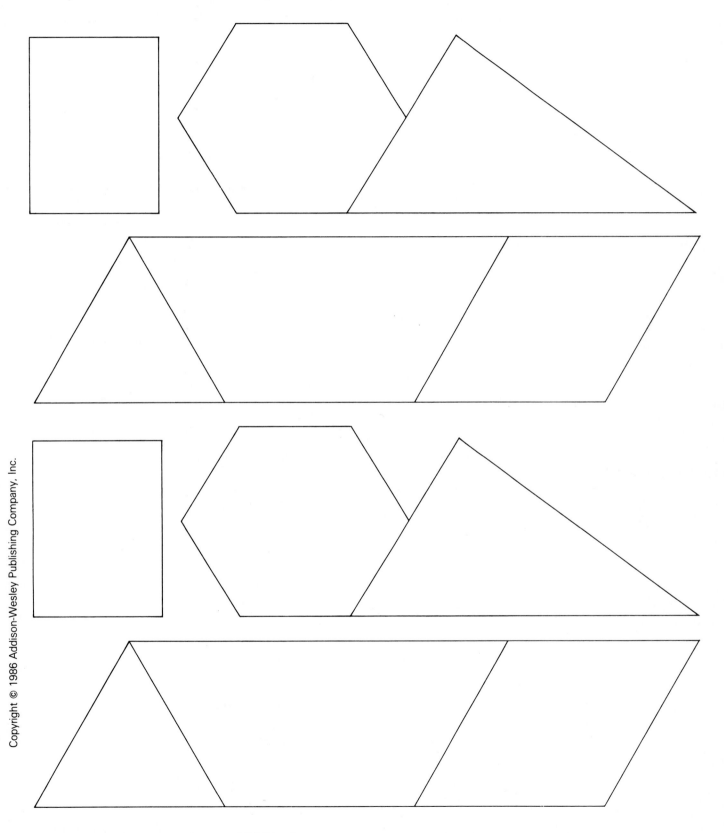

Reptiles I

Subdivide these figures into four congruent reptiles. Remember: Each small piece must be *similar* to the original.

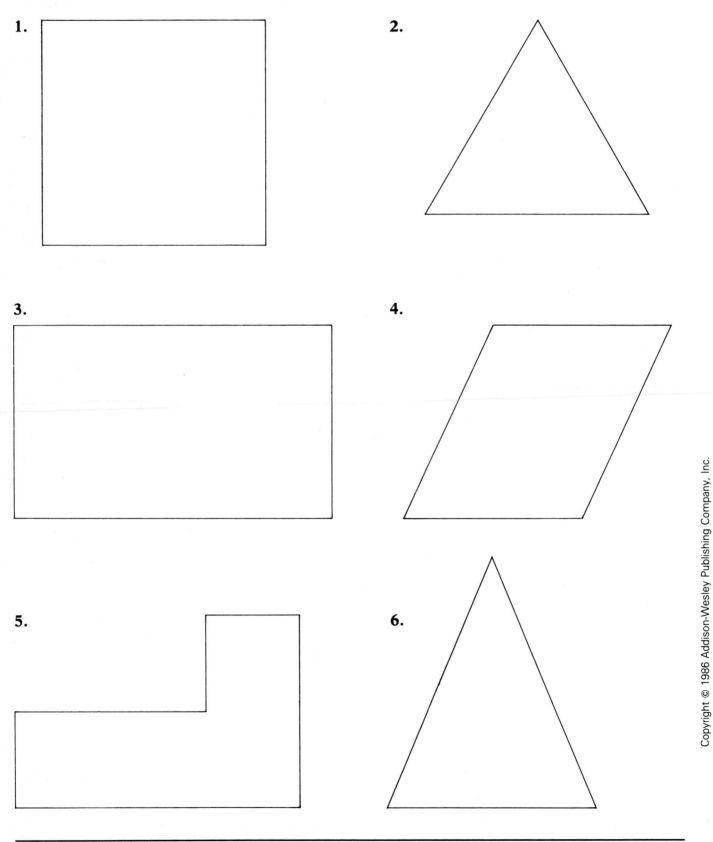

1.

2.

3.

4.

5.

6.

Worksheet 4-1

Reptiles II

Subdivide these figures into four congruent reptiles.

7.

8.

9.

10.

11.

12.

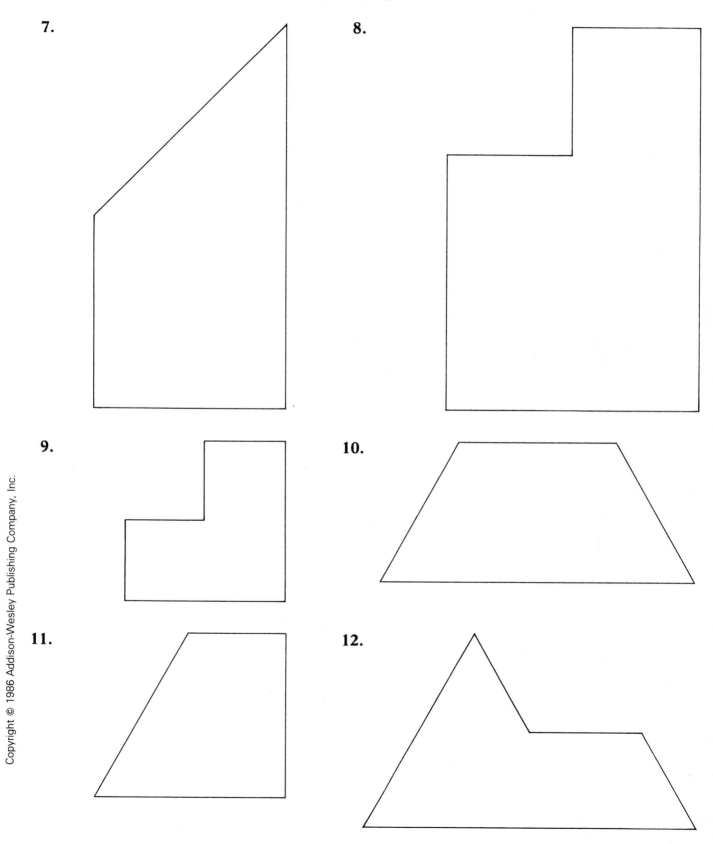

Super Reptiles

Cut along the lines.

Place the four Reptiles marked 1 on the reptile marked 2 so that they exactly cover.

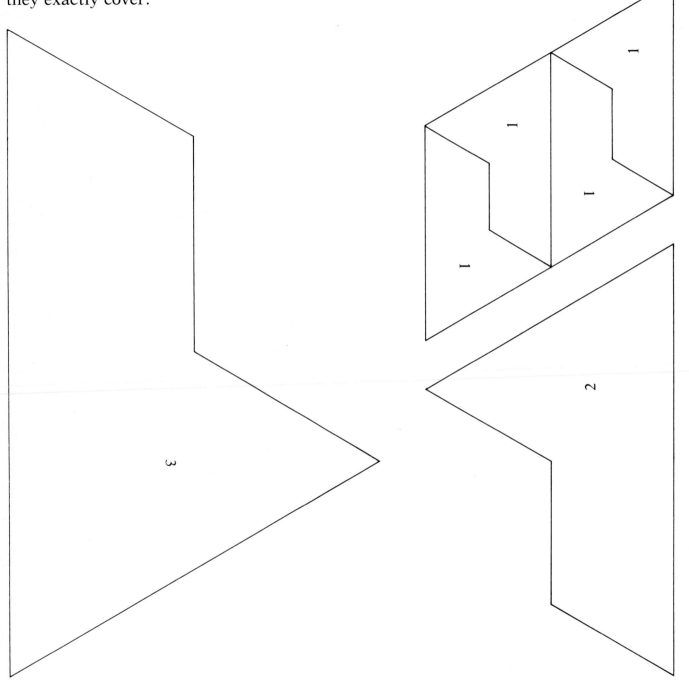

Extra Challenge
Find how many of Reptile 1 it takes to cover Reptile 3 exactly.

Worksheet 4-3

Are They Similar?

For each pair of figures show why they are or are not similar. You should check corresponding angles and ratios of corresponding sides to see that they are or are not equal.

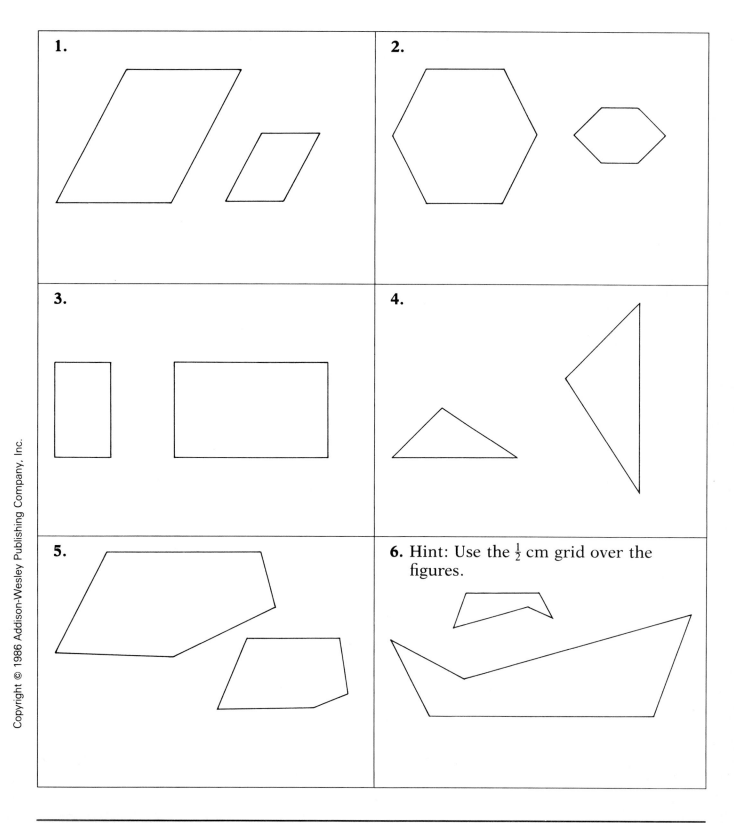

1.

2.

3.

4.

5.

6. Hint: Use the $\frac{1}{2}$ cm grid over the figures.

Activity 5

SIMILAR TRIANGLES

OVERVIEW

This activity is similar to Activity 3, Similar Rectangles. Students are asked to find, from among several right triangles, a right triangle that is not similar to the other triangles. Students use the transparent grid paper from Activity 3 to measure lengths of sides and area of triangles. They might use the strategy of finding ratios of corresponding sides and checking for equal ratios. After the similar triangles are determined, students are asked to find the scale factor relative to one of the given triangles. The *scale factor* is a number that describes the relative size from an object to its enlargement (or contraction). It is the number used to multiply the original dimensions to get the dimensions of the new similar figure. Students then verify that in similar triangles the perimeters grow by the scale factor, and the area grows by the square of the scale factor.

Students use "two triangles are similar" to find missing lengths. This prepares the students for a later activity involving similar triangles to find the height of a flagpole or other inaccessible objects.

Goals for students

1. Use the property that ratios of corresponding sides in two similar figures are equal to determine which triangles are similar.
2. Use the fact that two triangles are similar to find missing lengths of sides in one of the triangles.
3. Find the scale factor in a pair of similar triangles.
4. Verify that perimeters in similar triangles grow by the scale factor.
5. Verify that areas of similar triangles grow by the square of the scale factor.

Materials

*Transparent grid paper (Materials 3-1).
*Right Triangles (Materials 5-1).

Worksheets

*5-1, Triangles: Which Are Similar?
*5-1, page 2, Chart For Similar Triangles.
 5-2, Testing Right Triangles.

Transparencies

Starred items should be made into transparencies.

TEACHER ACTION	TEACHER TALK	EXPECTED RESPONSE
Review.	How do we determine when two rectangles are similar?	When the ratios of their sides are equal.
Demonstrate by cutting a paper.	If we cut a rectangle along a diagonal, we get two congruent right triangles.	
Ask.	What do we mean by congruent?	Same size and shape; this means that lengths are equal and angles are equal.
	What do we mean by right triangle?	One with a right angle.
	Do you remember the name for the sides of a right triangle?	
	When would two right triangles be similar?	When the ratios of their legs are equal.
	Let's check this out to see whether we all agree.	
Put a transparency of Materials 5-1 on the overhead. Expose the top two triangles.	This is a pair of similar right triangles.	
	Let's check the ratio of the short leg to the long leg. Are they equal?	Yes. $\dfrac{\text{Short leg}}{\text{Long leg}} = \dfrac{2}{3} = \dfrac{6}{9}$
Superimpose a grid to show that triangles are half of a rectangle.		

Activity 5 *Launch*

TEACHER ACTION	TEACHER TALK	EXPECTED RESPONSE
Expose the second two triangles.	These are similar right triangles. Can you find the missing length? How?	Yes; check ratios—the ratios must be equivalent. $$\frac{5}{15} = \frac{1}{3}; \frac{?}{6} = \frac{1}{3}$$ The answer is 2.
Pass out both pages of Worksheet 5-1, Triangles: **Which are Similar?** and Chart for Similar Triangles. Display a transparency of Worksheet 5-1, page 2. Review how to find lengths of the sides of right triangles using the grid paper as a ruler. Fill in the first row of data in the first three columns. Be sure to use the edge of the grid to measure length. No rulers!	We want to see if all of these triangles belong to the same family. Let's do triangle 1 together.	
	Measure the other seven triangles to see if they all belong. Fill out only the left part of the chart, sides and ratios.	

Sides			Ratio
a	b	c	$\frac{a}{b}$
9 units	12 units	15 units	$\frac{9}{12} = \frac{3}{4}$

Activity 5 *Explore*

OBSERVATIONS

If the grids are placed on top of the figures, students will miscount the length. They will count this length as 2 or $2\frac{1}{2}$.

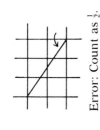

Error: Count as $\frac{1}{2}$.

As an extra challenge, introduce more advanced students to the Pythagorean relationship for right triangles.

POSSIBLE RESPONSES

We are using the lines on the grid to measure length, not the squares. Place the grid outside the figure for length.

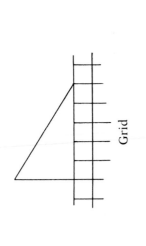

Grid

$a^2 + b^2 = c^2$

$3^2 + 4^2 = 5^2$

$9 + 16 = 25$

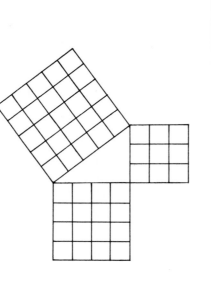

Activity 5 *Summarize*

TEACHER ACTION	TEACHER TALK						EXPECTED RESPONSE

TEACHER ACTION

Display a transparency of page 2 of Worksheet 5-1, Chart for Similar Triangles, on the overhead. Collect answers from class for the sides and ratios.

If students have trouble with triangle 8, give them the length of b. Tell the students it is similar and let them figure out the length of a and c.

TEACHER TALK

Sides					Ratio
Triangle	a	b	c		$\frac{a}{b}$
1	9 units	12 units	15 units		$\frac{9}{12} = \frac{3}{4}$
2	15	20	25		$\frac{15}{20} = \frac{3}{4}$
3	18	24	30		$\frac{18}{24} = \frac{3}{4}$
4	10	14	$17\frac{1}{2}$		$\frac{10}{14} = \frac{5}{7}$
5	3	4	5		$\frac{3}{4}$
6	6	8	10		$\frac{6}{8} = \frac{3}{4}$
7	12	16	20		$\frac{12}{16} = \frac{3}{4}$
8	$4\frac{1}{2}$	6	$7\frac{1}{2}$		$\frac{4\frac{1}{2}}{6} = \frac{3}{4}$

Ask.

Which triangle is not similar to the others?

How can you tell?

If we have a right triangle whose legs are 300 and 400, could this triangle be similar to the above triangles? Why?

Compared to triangle 5 what would be the scale factor for a similar triangle with legs 300 and 400?

EXPECTED RESPONSE

4

The ratio of legs is not the same.

Yes, ratio of the legs is $\frac{300}{400} = \frac{3}{4}$.

100.

Activity 5 *Summarize*

TEACHER ACTION

Fill in Scale Factor column. If the students have trouble with scale factor, ask, "What do I multiply the legs of triangle 5 by to get triangle 3?" "What do I multiply the legs of triangle 5 by to get triangle 2?" etc.

Superimpose the grid on one triangle as an example if necessary.

Give class time to do the chart.

Gather data.

TEACHER TALK

First mark out triangle 4 on the right side of your chart because it does not belong to the family of similar triangles.

Let's find the scale factor for each triangle in terms of the smallest similar triangle, which is triangle 5.

Now we need to find the perimeter and area of the family of similar right triangles.

Remember that *perimeter* is the distance around the figure. We can find perimeter by adding the lengths of all the sides of the figure.

The area of a right triangle is the number of unit squares needed to exactly cover the figure. It is easy to find the area if you think of each right triangle as half of a rectangle.

Fill in the perimeter and area columns of Worksheet 5-1, page 2, Chart for Similar Triangles.

Let's fill in the column for perimeters.

What pattern do you see in the perimeter growth?

EXPECTED RESPONSE

Scale Factor	Perimeter	Area
3	36 units	54 sq. units
5	60	150
6	72	216
X	X	X
1	12	6
2	24	24
4	48	96
$1\frac{1}{2}$	18	$13\frac{1}{2}$

If the sides are doubled the perimeter is doubled.

Activity 5 *Summarize*

TEACHER ACTION	TEACHER TALK	EXPECTED RESPONSE
Ask.	How do the perimeters in two similar triangles compare?	It grows by the scale factor.
Take several examples.		In triangle 1 the scale factor is 3 and the perimeter is 36, which is 3 times the perimeter of triangle 5.

Triangle 5   *Triangle 1*

	scale factor	
	3	

$P = 12$ $P = 36 = 3 \times 12$

Gather data:	Now let's fill in the column for areas.	
Ask.	How do the areas compare in similar right triangles?	Area grows by the square of the scale factor.

$A = 6$ (triangle with legs 3 and 4)

$A = 54 = 3 \times 3 \times 6$ (triangle with legs 9 and 12)

Activity 5 *Summarize*

TEACHER ACTION	TEACHER TALK	EXPECTED RESPONSE
Draw on the board or overhead.	If a right triangle has legs 2 and 3, describe a triangle that is similar to it.	Various answers. It will have legs that are 4 and 6, 6 and 9, 8 and 12, 20 and 30, and so on; 1 and $1\frac{1}{2}$, $\frac{2}{3}$ and 1, and so on.
Ask.	If the *area* of a right triangle is 9 square units, what would be the area of a similar right triangle with a scale factor of 3 between the triangles?	81 square units, or 9×3^2.
Pass out Worksheet 5-2, Testing Right Triangles. This may be given as homework.	Worksheet 5-2 asks you to test some right triangles to see if they are similar and to find the missing side in some pairs of similar right triangles.	

Right Triangles

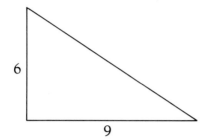

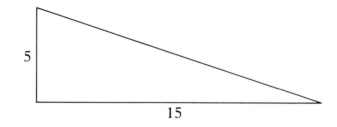

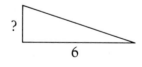

Triangles: Which Are Similar?

Measure the triangles with your grid. Record your information on the chart on Worksheet 5-1, page 2. Which triangles are similar?

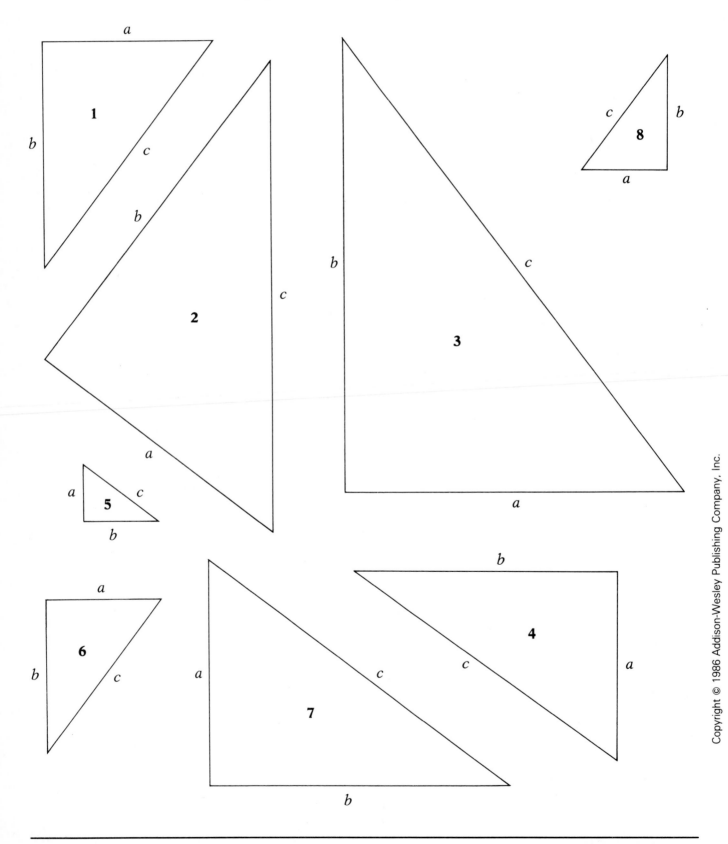

Worksheet 5-1

Chart for Similar Triangles

Which triangles are similar?

How do similar right triangles grow? Use triangle 5 to find the scale factor for the other similar triangles.

Describe how you can test two right triangles to see if they are similar.

| Triangle | Sides | | | Ratio | Scale Factor | Perimeter | Area |
	a	b	c	$\frac{a}{b}$			
1							
2							
3							
4							
5					1		
6							
7							
8							

Testing Right Triangles

A. Without measuring, tell which pairs are similar. _____

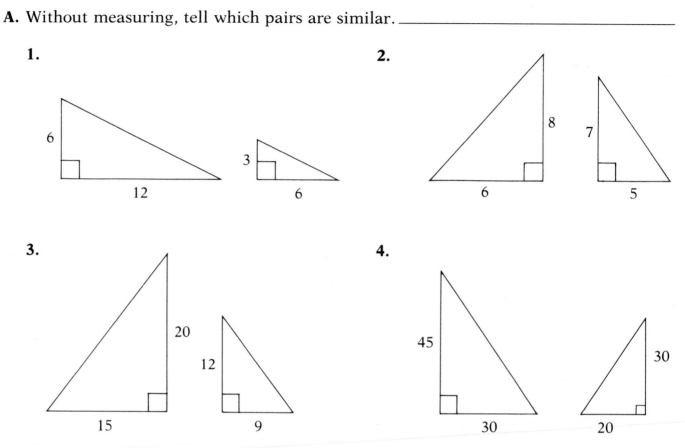

1.

6

12

3

6

2.

8

6

7

5

3.

20

15

12

9

4.

45

30

30

20

B. Each pair of right triangles is similar. Find the missing measurement.

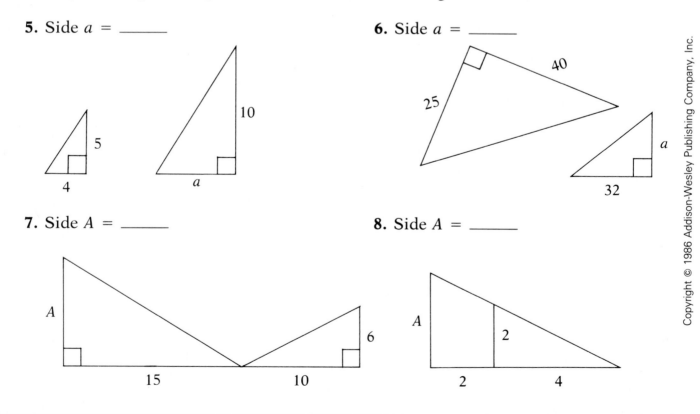

5. Side a = _____

5

4

10

a

6. Side a = _____

40

25

a

32

7. Side A = _____

A

15

10

6

8. Side A = _____

A

2

2

4

Optional Computer Extension for Testing Right Triangles

Applesoft Basic

Note: A teacher-prepared worksheet of right triangles to be measured
        is needed. Maximum legs: 17×22 cm

```
10    HOME : REM    "TESTING FOR SIMILAR RIGHT TRIANGLES USING RATIOS(DECI
      MAL AND FRACTIONAL) AND NESTING -FOR MSMP BY M. BJORNHOLM"
20    LOMEN: 16384: REM    "A TEACHER PREPARED WORKSHEET OF RIGHT TRIANGLES
      TO BE MEASURED IS NEEDED FOR EACH OF THE TWO SECTIONS."
150   GOSUB 40000
170   VTAB 22: HTAB 2: PRINT "TESTING FOR SIMILAR RIGHT TRIANGLES": FOR T
      = 1 TO 2000: NEXT T
190   HGR : TEXT : HOME : REM  "END LOGO -> START MENU"
200   VTAB 2: PRINT "YOU HAVE A CHOICE OF TWO METHODS TO TESTFOR SIMILARI
      TY OF RIGHT TRIANGLES. BOTH REQUIRE YOU TO CAREFULLY MEASURE TRI-
      ANGLES FROM YOUR WORKSHEET. YOU THEN IN-PUT YOUR DATA TO THE COMPUTE
      R WHEN RE-  QUESTED."
210   PRINT : PRINT "RATIO IS A PROGRAM THAT TESTS FOR SIMI- LARITY BY CO
      MPARING RATIOS OF THE TWO    LEGS OF RIGHT TRIANGLES. THE LEGS FORM
      THE RIGHT ANGLE."
220   PRINT : PRINT "NESTING INVOLVES PUTTING ONE TRIANGLE ONANOTHER WITH
      THE RIGHT ANGLES AT THE     SAME LOCATION TO TEST FOR SIMILARITY."
230   PRINT : PRINT "ALWAYS PRESS THE ";: INVERSE : PRINT "RETURN";: NORMAL
      : PRINT " KEY AFTER YOU   ANSWER A QUESTION."
240   GOSUB 50000
250   VTAB 5: PRINT "TYPE THE LETTER OF THE ITEM YOU WISH.": PRINT
260   HTAB 6: PRINT "A     RATIO": PRINT
270   HTAB 6: PRINT "B       NESTING": PRINT
280   HTAB 6: PRINT "C       I AM FINISHED": PRINT
290   PRINT "YOUR LETTER";: INPUT A$
300   IF A$ = "A" THEN 1000
310   IF A$ = "B" THEN 2000
320   IF A$ = "C" THEN 340
330   PRINT : PRINT "PRESS ONLY 'A', 'B', OR 'C' PLEASE!": GOTO 250
340   HOME : VTAB 10: PRINT "THANK YOU FOR WORKING WITH ME. I HOPE    YOU
      DID WELL AND LEARNED SOMETHING.";: END
1000  FOR T = 1 TO 1000: NEXT T
1010  HOME : REM  "RATIO METHOD FOR FINDING SIMILAR TRIANGLES. USE TEACH
      ER PREPARED WORKSHEET OF RIGHT TRIANGLES."
1020  GOSUB 40000
1030  VTAB 22: COLOR= 15: PRINT "THIS IS A PROGRAM USING RATIOS TO TEST
      WHETHER OR NOT TRIANGLES ARE IN THE SAMEFAMILY.       PRESS ANY KEY
      TO CONTINUE";: GET Z$: HGR : TEXT : HOME
1050  PRINT "TO DETERMINE IF TWO RIGHT TRIANGLES ARE SIMILAR FIND THE RA
      TIO OF THE SHORT LEG TO THE LONG LEG FOR EACH FIGURE. IF THE RATIOS
      ARE THE SAME THEN THE TRIANGLES  ARE SIMILAR, IN THE SAME FAMILY.":
      PRINT
```

```
1055    PRINT "AN EXAMPLE: THE LEGS OF ONE TRIANGLE ARE6 AND 8 WITH A HYPO
        TENUSE OF 10 WHILE    THE LEGS OF ANOTHER TRIANGLE ARE 9 AND  12 WITH
        A HYPOTENUSE OF 15. THE RATIO   6/8 = .75. THE RATIO OF 9/12 = .75.

1056    PRINT "THE FRACTIONS 6/8 AND 9/12 ARE EQUIVA-  LENT. AS THE RATIOS
        ARE THE SAME THE     TRIANGLES ARE SIMILAR."
1070    PRINT "WITH YOUR TRANSPARENT GRID MEASURE THE  LEGS OF EACH RIGHT
        TRIANGLE. GIVE THE    NUMERICAL VALUE ONLY, NO UNITS.": PRINT
1075    PRINT "IF THE LEGS OF A TRIANGLE ARE THE SAME  LENGTH GIVE THAT LE
        NGTH FOR BOTH SHORT  AND LONG LEG REQUESTS."
1080    GOSUB 50000
1090    HOME
1100    PRINT : PRINT "WHAT IS THE LENGTH OF THE SHORT LEG OF  THE FIRST T
        RIANGLE";: INPUT S
1110    PRINT : PRINT "WHAT IS THE LENGTH OF THE LONG LEG OF   THE FIRST T
        RIANGLE";: INPUT L
1120    PRINT : PRINT "WHAT IS THE LENGTH OF THE SHORT LEG OF   THE SECOND
        TRIANGLE";: INPUT T
1130    PRINT : PRINT "WHAT IS THE LENGTH OF THE LONG LEG OF    THE SECOND
        TRIANGLE";: INPUT M
1140    IF S > L THEN 1180: REM  "ERROR CONTROL"
1150    IF T > M THEN 1180: REM  "    LOOP"
1160    PRINT : PRINT : INVERSE : PRINT "WRITE";: NORMAL : PRINT " FRACTIO
        NS OF THE SHORT SIDE OVER  THE LONG SIDE AND DECIDE WHETHER OR NOT T
        HE FRACTIONS ARE EQUAL. THEN CONTINUE.": GOTO 1200
1180    PRINT : PRINT : PRINT : PRINT "YOU GAVE AN INCORRECT MEASUREMENT O
        R IN-PUT MEASUREMENT(S) IN THE WRONG PLACE.  TRY AGAIN."
1190    VTAB 22: HTAB 9: PRINT "PRESS ANY KEY TO CONTINUE": GET Z$: GOTO 1
        090
1200    GOSUB 50000
1210    PRINT "THE LEGS OF THE FIRST TRIANGLE ARE        ";S;" AND ";L;". ";
        S;"/";L;" HAS A RATIO OF ";S / L;".": PRINT
1220    PRINT "THE LEGS OF THE SECOND TRIANGLE ARE       ";T;" AND ";M;". ";
        T;"/";M;" HAS A RATIO OF ";T / M;".": PRINT
1230    IF T / M = S / L THEN 1270
1240    PRINT S;"/";L;" DOES NOT EQUAL ";T;"/";M;"."
1250    PRINT "THE TRIANGLES ARE NOT SIMILAR.": GOTO 1290
1270    PRINT S;"/";L;" EQUALS ";T;"/";M;"."
1280    PRINT "THE TRIANGLES ARE SIMILAR."
1290    PRINT : PRINT "DO YOU WISH TO TEST TWO MORE TRIANGLES  FOR SIMILAR
        ITY";: INPUT A$
1300    IF A$ = "Y" THEN 1090
1310    IF A$ = "N" THEN  HOME : GOTO 250
1320    HOME : VTAB 10: HTAB 6: PRINT "PLEASE PRESS 'Y' OR 'N' ONLY.";: INPUT
        A$: GOTO 1300
2000    REM  "NESTING: TEACHER PREPARE RIGHT TRIANGLE WORKSHEETS. MEASUREM
        ENT=NUMBER 1/2 CM UNITS. MAXIMUM LEG LENGTHS=19 AND 22."
2025    HGR : HCOLOR= 3
2030    HPLOT 170,159 TO 270,0: HPLOT  TO 270,159: HPLOT  TO 170,159
2040    HPLOT 0,100 TO 133,100: HPLOT  TO 0,153: HPLOT  TO 0,100
```

```
2050   VTAB 22: HTAB 3: PRINT "TWO RIGHT TRIANGLES TO BE NESTED"
2060   GOSUB 50000
2070   HGR : HCOLOR= 3
2080   HPLOT 0,0 TO 200,0: HPLOT  TO 0,80: HPLOT  TO 0,0
2085   FOR T = 1 TO 300: NEXT T
2090   HPLOT 0,0 TO 133,0: HPLOT  TO 0,53: HPLOT  TO 0,0
2120   VTAB 21: PRINT "THE TRIANGLES HAVE BEEN NESTED IN THE   RIGHT ANGL
       E. ";: INVERSE : PRINT "HYPOTENUSES";: NORMAL : PRINT " ARE PARALLEL
       SOTHE TRIANGLES ARE SIMILAR."
2128   VTAB 24: HTAB 9: PRINT "PRESS ANY KEY TO CONTINUE";
2130   GET Z$: HOME : VTAB 21: PRINT "THE TRIANGLES WILL ALWAYS BE NESTED
       IN  THE UPPER LEFT CORNER FOR TESTING."
2140   VTAB: 24: HTAB 9: PRINT "PRESS ANY KEY TO CONTINUE";
2145   GET Z$: HGR : TEXT : HOME
2150   PRINT "WITH YOUR TRANSPARENT GRID MEASURE THE  NUMBER OF UNITS OF
       LENGTH FOR THE SHORT AND THE LONG LEG FOR THE RIGHT TRIANGLESON YOUR
       WORKSHEET.": PRINT
2160   PRINT "FOLLOW THE DIRECTIONS TO HAVE THE RE-   SULTS OF THE TEST F
       OR NESTED TRIANGLES  SHOWN ON THE SCREEN."
2165   PRINT : PRINT "IF THE HYPOTENUSES ARE TOO CLOSE TO     BEING PARAL
       LEL TO MAKE A GOOD DECISION  USE THE RATIO METHOD TEST TO CHECK."
2170   GOSUB 50000
2180   PRINT "WHAT IS THE LENGTH OF THE SHORT LEG OF ATRIANGLE";: INPUT S
       : PRINT
2190   PRINT "HHAT IS THE LENGTH OF THE LONG LEG OF    THE TRIANGLE";: INPUT
       L: PRINT
2200   PRINT "WHAT IS THE LENGTH OF THE SHORT LEG OF  ANOTHER TRIANGLE";:
       INPUT T: PRINT
2210   PRINT "WHAT IS THE LENGTH OF THE LONG LEG OF    THIS TRIANGLE";: INPUT
       M: PRINT
2224   IF S > 19 OR T > 19 THEN 2260
2226   IF L > 22 OR M > 22 THEN 2260
2227   IF S > L THEN 2240: REM  "ERROR CONTROL"
2228   IF T > M THEN 2240: REM  "    LOOP"
2230   GOTO 2280
2240   PRINT : PRINT : PRINT "YOU GAVE A SHORT LEG OR A LONG LEG      MEA
       SUREMENT IN THE WRONG LOCATION.       TRY AGAIN."
2250   GOTO 2170
2260   PRINT : PRINT : PRINT "YOU GAVE TOO LARGE A MEASUREMENT.      MEA
       SURE AND INPUT AGAIN."
2270   GOTO 2170
2280   PRINT "THE LEGS OF THE TRIANGLE ARE ";S;" BY ";L;" AND ";T;" BY ";
       M:"."
2290   VTAB 17: INVERSE : PRINT "DECIDE";: NORMAL : PRINT " IF THE TRIANG
       LES ARE SIMILAR."
2300   GOSUB 50000
2310   HOME: HGR : HCOLOR= 3
2320   HPLOT 0,0 TO  INT (12.5 * M),0: HPLOT  TO 0,10 * T: HPLOT  TO 0,0
2325   FOR T = 1 TO 300: NEXT T
2330   HPLOT 0,0 TO  INT (12.5 * L),0: HPLOT  TO 0,10 * S: HPLOT  TO 0,0
```

```
2340   VTAB 22: PRINT "ARE THE HYPOTENUSES PARALLEL?"
2350   VTAB 24: PRINT "ARE THE TRIANGLES SIMILAR?"
2360   FOR T = 1 TO 2500: NEXT T
2370   VTAB 22: PRINT
2380   VTAB 24: HTAB 9: PRINT "PRESS ANY KEY TO CONTINUE";: GET Z$: HGR :
       TEXT
2390   HOME : VTAB 10: PRINT "DO YOU WISH TO TEST MORE - `Y` OR `N`";: INPUT
   A$
2400   IF A$ = "Y" THEN 2170
2410   IF A$ = "N" THEN  HOME : GOTO 250
2420   HOME: VTAB 6: HTAB 6: PRINT "PLEASE PRESS `Y` OR `N` WHEN ASKED."

2430   VTAB 22: HTAB 9: PRINT "PRESS ANY KEY TO CONTINUE";: GET Z$: GOTO
       2390
40000   GR : COLOR= 7: HLIN 5,0 AT 15: VLIN 15,20 AT 0: PLOT 1,19: PLOT 2
       ,18: PLOT 3,17: PLOT 4,16
40005   COLOR= 9: HLIN 24,39 AT 5: VLIN 5,20 AT 39: PLOT 38,19: PLOT 37,1
       8: PLOT 36,17: PLOT 35,16: PLOT 34,15: PLOT 33,14: PLOT 32,13: PLOT
       31,12: PLOT 30,11: PLOT 29,10: PLOT 28,9: PLOT 27,8: PLOT 26,7: PLOT
       25,6
40010   COLOR= 5: HLIN 10,20 AT 35: VLIN 35,25 AT 20: PLOT 19,26: PLOT 18
       ,27: PLOT 17,28: PLOT 16,29: PLOT 15,30: PLOT 14,31: PLOT 13,32: PLOT
       12,33: PLOT 11,34: RETURN
50000   VTAB 24: HTAB 9: PRINT "PRESS ANY KEY TO CONTINUE";: GET Z$: HOME
       : RETURN
```

QUICK TESTS FOR SIMILARITY OF RECTANGLES AND TRIANGLES

OVERVIEW

In this activity we explore sets of similar rectangles and triangles, which we call families. If two figures are similar, they belong to the same family, and vice versa. We test for similar rectangles by finding the ratio of the short side to the long side. If the ratios for two rectangles are equal, the rectangles are similar. If the ratios of corresponding sides of triangles are equal, the triangles are similar.

As we complete this activity the students discover that if two rectangles are similar, we can nest them at one corner and their diagonals will coincide. Therefore, an entire family of similar rectangles will share the same diagonal when nested, for example, at their lower left corner. The test for triangles is to nest at a corresponding angle and observe that the opposite sides are parallel.

Goals for students

1. Learn that similar rectangles and triangles can be thought of in families.
2. Learn that when similar rectangles are nested at a common corner, their diagonals coincide.
3. Learn that when similar triangles are nested at a corresponding angle, the opposite sides are parallel.

Materials

Straight edge.

Worksheets

6-1, Grid for Rectangles.
*6-1, page 2, Rectangles.
6-2, Triangles.
6-3, Which Are Similar?
6-4, Similar Rectangles.

Transparencies

Starred item should be made into transparency.

QUICK TESTS FOR SIMILARITY OF RECTANGLES AND TRIANGLES

TEACHER ACTION	TEACHER TALK	EXPECTED RESPONSE
Students need Worksheet 6-1, Grid for Rectangles.		
Make a table to show the answers on the overhead.	Look carefully at each of the rectangles A, B, C, D, E, F, G, and H.	
	How many H's would it take to cover G?	2
	How many H's would it take to cover F?	4
	How many H's would it take to cover E?	8
	How many H's would it take to cover D?	16
	How many H's would it take to cover C?	32
	How many H's would it take to cover B?	64
	How many H's would it take to cover A?	128
	Now, let's look at the rectangles in terms of A.	
	How many A's would it take to cover A?	1
	How many A's would it take to cover B?	$\frac{1}{2}$
	How many A's would it take to cover C?	$\frac{1}{4}$
	How many A's would it take to cover D?	$\frac{1}{8}$
	How many A's would it take to cover E?	$\frac{1}{16}$
	How many A's would it take to cover F?	$\frac{1}{32}$
	How many A's would it take to cover G?	$\frac{1}{64}$
	How many A's would it take to cover H?	$\frac{1}{128}$
	Good. So how are the rectangles constructed starting with A?	Cut the grid in half.
	Then how do we get B?	Cut half of the grid in half.
	So each rectangle is half the one before it. For example, E is $\frac{1}{2}$ of D; G is $\frac{1}{2}$ of F, and so on.	

	A	B	C	D	E	F	G	H
No. of H's Needed to cover	128	64	32	16	8	4	2	1
No. of A's Needed to cover	1	$\frac{1}{2}$	$\frac{1}{4}$	$\frac{1}{8}$	$\frac{1}{16}$	$\frac{1}{32}$	$\frac{1}{64}$	$\frac{1}{128}$

Activity 6 *Launch*

TEACHER ACTION	TEACHER TALK	EXPECTED RESPONSE
Now cut out each of the rectangles A, B, C, D, E, F, G, H, and I.	What are the dimensions of A?	20×32 units.
	What is the short side?	20 units
	What is the long side?	32 units
	What is the ratio of the short side to the long side?	$\frac{20}{32} = \frac{5}{8}$
	Give me the dimensions of another rectangle similar to A.	Various answers.
	Describe how you can test two rectangles for similarity.	The ratios, $\frac{shortside}{longside}$, must be equal.
	Label your A like this and draw the diagonals.	

Ask.	What are the dimensions of B?	16×20 units.
	What is the short side?	16 units.
	What is the long side?	20 units.
	What is the ratio of the short side to the long side?	$\frac{16}{20} = \frac{4}{5}$
Put on the board.	Take the B and label it like this. Put this B aside with your A.	
	Draw the diagonals.	
	Label the dimensions of each of your remaining figures and draw the diagonals.	

81

TEACHER ACTION	TEACHER TALK	EXPECTED RESPONSE
Give directions. Illustrate.	Nest your collection of rectangles in order of size with their lower left corners aligned and the long sides down.	
Ask.	What do you notice? This gives us another test for similar rectangles. If we nest rectangles at a corner with the long sides down, one of the sets of diagonals will fall on a straight line.	The diagonals for the two families align.
Students need Worksheet 6-1, page 2, Rectangles. Give directions.	Fill in the chart in number 1 with data from your rectangles.	

Activity 6 *Summarize*

TEACHER ACTION	**TEACHER TALK**	**EXPECTED RESPONSE**

Ask.

Let the students give a piece of information for any part of the chart. Let patterns come from the students. Don't force the results to be given in order A–H.

Give me a piece of information for the first three columns of the chart. Record the data in your chart as I record them at the board.

Rectangle	A	B	C	D	E	F	G	H
Short Side	20 units	16 units	10 units	8 units	5 units	4 units	$2\frac{1}{2}$ units	2 units
Long Side	32 units	20 units	16 units	10 units	8 units	5 units	4 units	$2\frac{1}{2}$ units
Ratio $\frac{short}{long}$	$\frac{5}{8}$	$\frac{4}{5}$	$\frac{5}{8}$	$\frac{4}{5}$	$\frac{5}{8}$	$\frac{4}{5}$	$\frac{5}{8}$	$\frac{4}{5}$

When the chart is completed, ask.

Some students may volunteer information while the chart is being filled out.

What patterns do you see?

The ratios alternate, $\frac{5}{8}$ then $\frac{4}{5}$.

Every other dimension for the short side is half.

A	C	E	G	B	D	F	H
20	10	5	$2\frac{1}{2}$	32	16	8	4

Activity 6 *Summarize*

TEACHER ACTION	TEACHER TALK	EXPECTED RESPONSE
Give directions. You can have the students fill this chart individually and then discuss, or you can do the chart as a class activity. If you do this as a class, fill in the table in a random order so that the patterns emerge only after the table begins to fill up.	Fill in the second chart on Worksheet 6-1, page 2.	

	Family I				Family II			
Rectangles	A	C	E	G	B	D	F	H
Area	640 sq. units	160 sq. units	40 sq. units	10 sq. units	320 sq. units	80 sq. units	20 sq. units	5 sq. units
Perimeter	104 units	52 units	26 units	13 units	72 units	36 units	18 units	9 units

TEACHER TALK	EXPECTED RESPONSE
When the table is completed, ask.	
What patterns do you see in the table? (Probe until students see patterns in perimeter and area)	Area gets small faster than perimeter.
Look at the family A, C, E, G. What patterns do you see in the perimeter?	104, 52, 26, 13; they are one half the previous number.
The number that tells you how the length of the sides of two similar figures relate is called the *scale factor*.	
The scale factor of A to C is one half, C to E is one half, and so on.	
What would be the perimeter of Rectangle I, if we continued to cut?	$\frac{13}{2}$ or $6\frac{1}{2}$
What about the areas of A, C, E, G?	640, 160, 40, 10; they are all one fourth the previous number.
What happens with B, D, F, H?	The same thing! Perimeter is one half of the previous one, area is one fourth of the previous one.
Do you see a pattern in the number of H's needed to cover each of the others?	Yes; the numbers double each time from H to A. Within a family the number quadruples from one to the next.
Look only at a family of similar rectangles—A, C, E, G or B, D, F, H. Now what is the pattern for the number of H's?	128, 32, 8, 2 or 64, 16, 4, 1. It is one fourth each time, the same as the area.

Activity 6 *Summarize*

TEACHER ACTION	TEACHER TALK	EXPECTED RESPONSE
You may want to continue the lesson.	Let's name all the rectangles in terms of E.	
	How many E's would it take to cover A?	16
	How many E's would it take to cover B?	8
	How many E's would it take to cover C?	4
	How many E's would it take to cover D?	2
	How many E's would it take to cover E?	1
	How many E's would it take to cover F?	$\frac{1}{2}$
	How many E's would it take to cover G?	$\frac{1}{4}$
	How many E's would it take to cover H?	$\frac{1}{8}$
	A, C, E and G are similar and the pattern for area is still one fourth.	
Summarize.	To summarize what we have seen: In similar figures, if the sides are twice as long, the area is four times as large or if the sides are half as long, the area is one fourth as large or the square of the scale factor, $(\frac{1}{2})^2 = \frac{1}{4}$.	
Have students work in pairs.	Let's check to see if a test like the diagonal test for rectangles works with right triangles.	
Pass out Worksheet 6-2, Triangles.	Cut out the triangles on Worksheet 6-2, Triangles.	
	Nest the triangles, largest on the bottom, at the right angle.	
	What do you observe?	The hypotenuses are parallel.

Activity 6 *Summarize*

TEACHER ACTION	TEACHER TALK	EXPECTED RESPONSE
	Nest at one of the acute angles, being careful to keep corresponding legs together.	
	What do you observe?	The legs are parallel.
	Are the two pairs of acute angles equal?	Yes.
Ask	What two tests do we have for determining whether right triangles are similar?	1. Ratios of the legs are equivalent. 2. When nested at a corresponding angle, the opposite sides are parallel.
Summarize. Draw an example on the board. 	We developed these tests for right triangles, but both tests are valid for testing the similarity of *any* two triangles if we are always careful to work with *corresponding* sides and if we remember to check that corresponding angles are equal.	
If students have difficulty seeing which sides correspond, superimpose triangles so that the marked angles coincide.	In the example, if the triangles are similar, the ratio $\frac{a}{c}$ in the first triangle will equal $\frac{a'}{c'}$ in the second.	
Assign Worksheet 6-3, Which Are Similar?		
Give out Worksheet 6-4, Similar Rectangles. (This may be assigned as homework.)	Draw the diagonals in each of the rectangles on Worksheet 6-4. Then cut them out and match them by their diagonals to find their similar families.	The apparent families are A, B, J; C, H; D, F, G; and E, I.

Grid for Rectangles

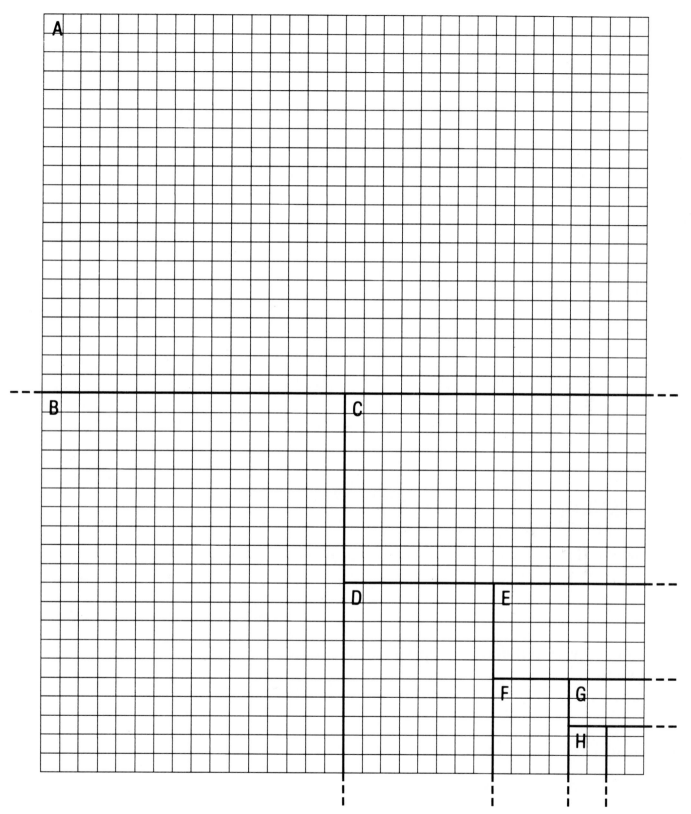

A

B   C

D   E

F   G

H

Rectangles

1. Record the data.

Rectangle	A	B	C	D	E	F	G	H
Short Side								
Long Side								
Ratio $\frac{short}{long}$								

2. Record the families of similar rectangles that you find in the chart.

Family I: _____

Family II: _____

3. What patterns do you see? _____

	Family I					Family II			
Rectangles	A	C	E	G		B	D	F	H
Area									
Perimeter									

4. How are the perimeters and areas related in each family of similar rectangles?

5. Describe every different test for similar rectangles that you have studied.

6. Consider the pair of rectangles A and E. What is the scale factor

from A to E? _____ What is the scale factor from E to A? _____

Triangles

Worksheet 6-2

Which Are Similar?

Cut out triangles A–G. Inside triangles 1, 2, 3, and 4, mark the letters of the triangles A–G that are similar to 1–4.

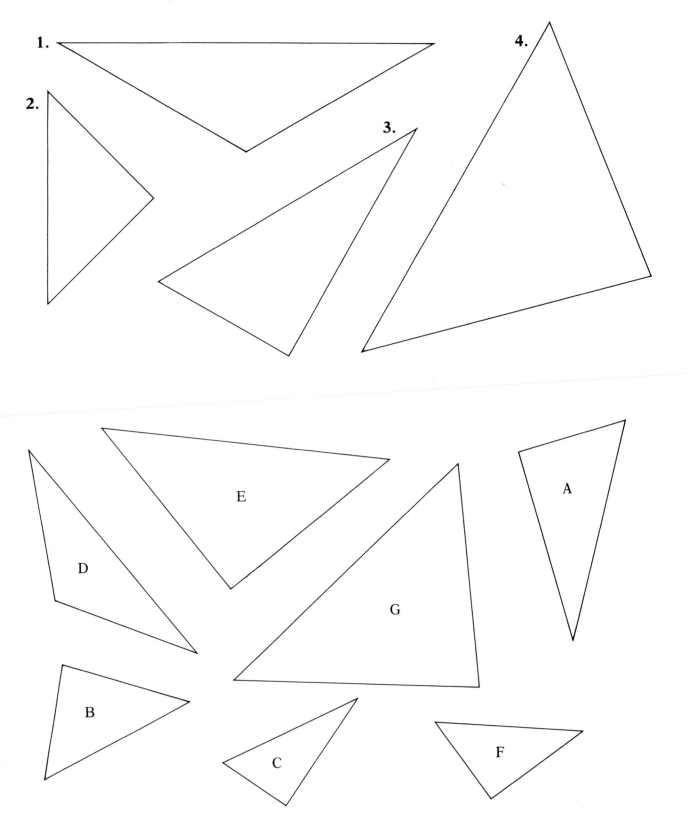

Worksheet 6-3

Similar Rectangles

Use the method of diagonals to sort these into similar rectangles.

1. Draw both diagonals in each rectangle.

2. Cut the rectangles out and find their similar families.

3. Make a record of your results.

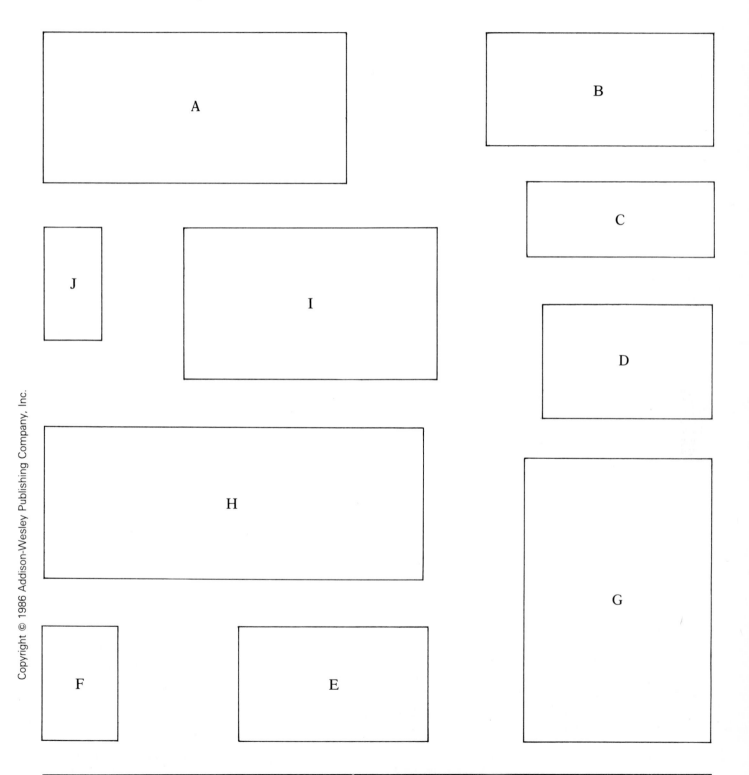

Optional Computer Extension for Testing Rectangles

Applesoft Basic

Note: Lines 40 to 330 may be omitted.

The maximum size on rectangles is 17 × 22 cm.

```
10   REM      "TESTING FOR SIMILARITY USING RATIOS (DECIMAL AND FRACTIONAL
     ) & NESTED RECTANGLES-FOR *MGMP* BY M. BJORNHOLM"
20   LOMEN: 16384: REM   "A WORKSHEET OF RECTANGLES TO BE MEASURED IS NEE
     DED FOR EACH OF THE TWO  SECTIONS IN THE PROGRAM."
40   HOME : REM  "LOGO-TITLE->SIMILARITY"
50   GR : COLOR= 3: HLIN 3,0 AT 15: VLIN 15,18 AT 0: HLIN 0,3 AT 18: VLIN
     18,22 AT 3: HLIN 3,0 AT 22
60   COLOR= 5: HLIN 6,8 AT 15: VLIN 16,21 AT 7: HLIN 6,8 AT 22
70   COLOR= 7: VLIN 22,15 AT 10: PLOT 11,16: PLOT 12,17: PLOT 13,16: VLIN
     15,22 AT 14
80   COLOR= 9: HLIN 16,18 AT 15: VLIN 16,21 AT 17: HLIN 16,18 AT 22
90   COLOR= 11: VLIN 15,22 AT 20: HLIN 21,22 AT 22
100   COLOR= 13: VLIN 22,17 AT 24: PLOT 25,16: PLOT 26,15: PLOT 27,16: VLIN
      17,22 AT 28: HLIN 25,27 AT 18
110   COLOR= 6: VLIN 22,15 AT 30: HLIN 31,33 AT 15: VLIN 16,18 AT 34: HLIN
      33,31 AT 19: PLOT 32,20: PLOT 33,21: PLOT 34,22
120   COLOR= 8: HLIN 37,39 AT 15: VLIN 16,18 AT 39: PLOT 38,18: VLIN 18,2
      0 AT 38: PLOT 38,22
130   FOR T = 1 TO 2500: NEXT T
140   HOME : REM  "LOGO-DESIGN->RANDOM RECTANGLES"
150   GR : FOR K = 1 TO 12
160   M = 35 *  RND (1):N = 35 *  RND (1)
170   P =  INT (6 *  RND (1)):R =  INT (6 *  RND (1))
180   C = 0
190   IF M < 4 THEN M = 4: IF M > 30 THEN M = 30
200   IF N < 7 THEN N = 7: IF N > 30 THEN N = 30
210   COLOR=  INT (15 *  RND (1) + 1)
220   Y = N - 4
230   FOR X = M - 4 TO M + P: PLOT X,Y: NEXT X
240   X = X - 1
250   FOR Y = N - 4 TO N + R: PLOT X,Y: NEXT Y
260   Y = Y - 1
270   FOR X = M + P TO M - 4 STEP  - 1: PLOT X,Y: NEXT X
280   X = X + 1
290   FOR Y = N + R TO N - 4 STEP  - 1: PLOT X,Y: NEXT Y
300   FOR T = 1 TO 900: NEXT T
310   NEXT K
320   VTAB 22: HTAB 6: PRINT "TESTING FOR SIMILAR RECTANGLES"
330   FOR T = 1 TO 2500: NEXT T
340   TEXT : HOME : REM  "END LOGO-START MENU"
350   VTAB 2: PRINT "YOU HAVE A CHOICE OF TWO METHODS TO TESTFOR SIMILARI
      TY OF RECTANGLES. BOTH RE-  QUIRE YOU TO CAREFULLY MEASURE THE REC-
      TANGLES FROM YOUR WORKSHEET. YOU THEN    INPUT YOUR DATA TO THE COMPU
      TER WHEN RE-QUESTED."
360    PRINT : PRINT "RATIO IS A PROGRAM THAT TESTS FOR SIMI- LARITY BY CO
      MPARING RATIOS OF THE SIDES OF RECTANGLES."
```

```
370   PRINT : PRINT "NESTING INVOLVES PUTTING ONE RECTANGLE  ON ANOTHER T
      O TEST FOR SIMILARITY."
380   PRINT : PRINT "ALWAYS PRESS THE ";: INVERSE : PRINT "RETURN";: NORMAL
      : PRINT " KEY AFTER
390   PRINT "YOU ANSWER A QUESTION."
400   VTAB 22: HTAB 12: PRINT "PRESS C TO CONTINUE";: GET Z$: HOME
410   HOME
420   VTAB 5: PRINT "TYPE THE LETTER OF THE ITEM YOU WISH.": PRINT
430   HTAB 6: PRINT "A     RATIO": PRINT
440   HTAB 6: PRINT "B     NESTING": PRINT
450   HTAB 6: PRINT "C     I AM FINISHED.": PRINT
460   PRINT "YOUR LETTER ";: INPUT A$
470   IF A$ = "A" THEN 520
480   IF A$ = "B" THEN 870
490   IF A$ = "C" THEN 510
500   PRINT : PRINT "CHOOSE A, B, OR C PLEASE!": GOTO 420
510   HOME : VTAB 10: PRINT "THANK YOU FOR WORKING WITH ME. I HOPE   YOU
      DID WELL AND LEARNED SOMETHING.": END
520   FOR T = 1 TO 1000: NEXT T: HOME
530   HOME : REM  "END MENU-START RATIO"
540   REM  "RATIO METHOD FOR FINDING SIMILAR REC-    TANGLES-WORKSHEET # 1
      "
550   GR : COLOR= 5: HLIN 5,15 AT 12: HLIN 5,15 AT 18: VLIN 13,17 AT 5: VLIN
      13,17 AT 15
560   COLOR= 8: HLIN 20,35 AT 12: HLIN 20,35 AT 21: VLIN 13,20 AT 20: VLIN
      13,20 AT 35
570   VTAB 22: COLOR= 15: PRINT "THIS IS A PROGRAM USING RATIOS TO TEST
      WHETHER OR NOT RECTANGLES ARE IN THE    SAME FAMILY.         PRESS C T
      O CONTINUE";
580   GET Z$: TEXT : HOME
590   PRINT "TO DETERMINE IF TWO RECTANGLES ARE SI-  MILAR FIND THE RATIO
      OF THE SHORT SIDE   TO THE LONG SIDE FOR EACH FIGURE. IF THERATIOS A
      RE THE SAME THE RECTANGLES ARE   SIMILAR, IN THE SAME FAMILY.": PRINT
600   PRINT "WITH YOUR TRANSPARENT GRID MEASURE THE   SHORT AND LONG SIDE
      OF EACH RECTANGLE.  GIVE THE NUMERICAL VALUE ONLY, NO UNITS."
610   VTAB 22: HTAB 12: PRINT "PRESS C TO CONTINUE";: GET Z$
620   HOME
630   PRINT "WHAT IS THE LENGTH OF THE SHORT SIDE OF THE FIRST RECTANGLE"
      ;: INPUT S
640   PRINT : PRINT "WHAT IS THE LENGTH OF THE LONG SIDE OF  THE FIRST RE
      CTANGLE";: INPUT L
650   PRINT : PRINT "WHAT IS THE LENGTH OF THE SHORT SIDE OF ANOTHER RECT
      ANGLE";: INPUT T
660   PRINT : PRINT "WHAT IS THE LENGTH OF THE LONG SIDE OF   THE RECTANGL
      E";: INPUT M
670   IF T > M THEN 710: IF S > L THEN 710: REM    "ERROR CONTROL LOOP"
690   PRINT : PRINT : INVERSE : PRINT "WRITE";: NORMAL : PRINT " FRACTION
      S OF THE SHORT SIDE OVER  THE LONG SIDE AND DECIDE WHETHER OR NOT TH
      E FRACTIONS ARE EQUAL - THEN CONTINUE."
```

```
700   GOTO 730
710   PRINT : PRINT : PRINT: PRINT "YOU GAVE AN INCORRECT MEASUREMENT OR
      IN-PUT ONE IN THE WRONG LOCATION. TRY AGAIN"
720   VTAB 24: HTAB 12: PRINT "PRESS C TO CONTINUE";: GET Z$: GOTO 620
730   VTAB 22: HTAB 12: PRINT "PRESS C TO CONTINUE";: GET Z$: HOME
740   PRINT "THE FIRST RECTANGLE IS ";S;" BY ";L;". IT HAS A RATIO OF ";S
      / L;".": PRINT
750   PRINT "THE NEXT RECTANGLE IS ";T;" BY ";M;". IT HAS A RATIO OF ";T /
      M;".": PRINT
760   IF T / M = S / L THEN 800
770   PRINT S;"/";L;" DOES NOT EQUAL ";T;"/";M;"."
780   PRINT "THE RECTANGLES ARE NOT SIMILAR."
790   GOTO 820
800   PRINT S;"/";L;" EQUALS ";T;"/";M;"."
810   PRINT "THE RECTANGLES ARE SIMILAR."
820   PRINT : PRINT "DO YOU WISH TO TEST TWO MORE RECTANGLES FOR SIMILARI
      TY - Y OR N";: INPUT A$
830   IF A$ = "Y" THEN 610
840   IF A$ = "N" THEN 410
850   HOME : PRINT : PRINT "PLEASE PRESS 'Y' OR 'N' ONLY."
860   VTAB 24: HTAB 12: PRINT "PRESS C TO CONTINUE";: GET Z$: HOME : GOTO
      820
870   HOME : REM  "END RATIO-START NESTING"
890   HGR : HCOLOR= 3: REM    "USE SHEET OF LETTERED RECTANGLES, MEASUREM
      ENT=NUMBER OF 1/2 CM UNITS , MAXIMUM=15×23 UNITS"
900   HPLOT 0,0 TO 49,0: HPLOT  TO 49,39: HPLOT  TO 0,39: HPLOT  TO 0,0:
      HPLOT 0,39 TO 49,0
910   HPLOT 100,129 TO 100,50: HPLOT  TO 199,50: HPLOT  TO 199,129: HPLOT
      TO 100,129: HPLOT  TO 199,50
920   VTAB 21: PRINT "DIFFERENT RECTANGLES WITH DIAGONALS"
930   VTAB 24: HTAB 12: PRINT "PRESS C TO CONTINUE";: GET Z$: HGR : HOME

940   HGR : HCOLOR= 3
950   HPLOT 100,129 TO 100,50: HPLOT  TO 199,50: HPLOT  TO 199,129: HPLOT
      TO 100,129: HPLOT  TO 199,50
960   HPLOT 100,129 TO 100,89: HPLOT  TO 149,89: HPLOT  TO 149,129: HPLOT
      TO 100,129: HPLOT  TO 149,89
970   VTAB 24: PRINT "THE RECTANGLES HAVE BEEN NESTED IN LOWERLEFT CORNER
      . DIAGONALS ALIGN SO THE RECTANGLES ARE SIMILAR. PRESS C TO CONTINUE
      ";
980   GET Z$: HGR : TEXT : HOME
990   PRINT "WITH YOUR TRANSPARENT GRID MEASURE THE  NUMBER OF UNITS OF L
      ENGTH FOR THE SHORT SIDE AND FOR THE LONG SIDE FOR EACH REC!TANGLE O
      N YOUR WORKSHEET.": PRINT
1000  PRINT "FOLLOW THE DIRECTIONS TO HAVE THE RE-    SULTS OF THE TEST F
      OR NESTED RECTANGLES SHOWN ON THE SCREEN."
1010  PRINT : PRINT "IF THE BASE IS SHORTER THAN THE HEIGHT  THE RECTANG
      LE WILL BE 'TIPPED' TO DRAW."
1020  VTAB 22: HTAB 12: PRINT "PRESS C TO CONTINUE";: GET Z$: HOME
1030  HGR : TEXT : HOME
```

```
1040   PRINT "WHAT IS THE LENGTH OF THE SHORT SIDE OF A RECTANGLE";: INPUT
       S: PRINT
1050   PRINT "WHAT IS THE LENGTH OF THE LONG SIDE OF  THE RECTANGLE";: INPUT
       L: PRINT
1060   PRINT "WHAT IS THE LENGTH OF THE SHORT SIDE OF ANOTHER RECTANGLE";
       : INPUT T: PRINT
1070   PRINT "WHAT IS THE LENGTH OF THE LONG SIDE OF  THIS RECTANGLE";: INPUT
       M: PRINT
1080   IF S > L THEN 1150: REM  "ERROR CONTROL LOOP"
1090   IF T > M THEN 1150
1100   IF S > 15 THEN 1180
1110   IF T > 15 THEN 1180
1120   IF L > 23 THEN 1180
1130   IF M > 23 THEN 1180
1140   GOTO 1210
1150   PRINT : PRINT : PRINT : PRINT "YOU GAVE A SHORT SIDE OR A LONG SID
       E    MEASUREMENT IN THE WRONG LOCATION.     TRY AGAIN."
1160   VTAB 22: HTAB 12: PRINT "PRESS C TO CONTINUE";
1170   GET Z$: GOTO 1030
1180   PRINT : PRINT : PRINT : PRINT "YOU GAVE TOO LARGE A MEASUREMENT.
       MEASURE AND INPUT AGAIN."
1190   VTAB 22: HTAB 12: PRINT "PRESS C TO CONTINUE";
1200   GET Z$: GOTO 1030
1210   PRINT "THE RECTANGLES ARE ";S;" BY ";L;" AND ";T;" BY ";M
1220   PRINT : PRINT : PRINT : INVERSE : PRINT "DECIDE";: NORMAL : PRINT
       " IF THE RECTANGLES ARE SIMILAR."
1230   VTAB 22: HTAB 12: PRINT "PRESS C TO CONTINUE";: GET Z$
1240   HOME : HGR : HCOLOR= 3
1250   HPLOT 0,159 TO 0,159 - 10 * S: HPLOT  TO 0 + 12 * L,159 - 10 * S:
       HPLOT  TO 0 + 12 * L,159: HPLOT  TO 0,159: HPLOT  TO 0 + 12 * L,159 -
       10 * S
1260   FOR Y = 1 TO 600: NEXT Y
1270   HPLOT 0,159 TO 0,159 - 10 * T: HPLOT  TO 0 + 12 * M,159 - 10 * T:
       HPLOT TO 0 + 12 * M,159: HPLOT  TO 0,159: HPLOT  TO 0 + 12 * M,159 -
       10 * T
1280   VTAB 24: PRINT "ARE THE RECTANGLES SIMILAR?"
1290   HTAB 12: PRINT "PRESS C TO CONTINUE";: GET Z$: HGR : TEXT
1300   HOME : VTAB 24: PRINT "DO YOU WISH TO TEST ANOTHER- 'Y' OR 'N'": INPUT
       A$
1310   IF A$ = "Y" THEN 1030
1320   IF A$ = "N" THEN 410
1330   HOME : VTAB 6: HTAB 6: PRINT "PLEASE PRESS 'Y' OR 'N' ONLY
1340   VTAB 22: HTAB 12: PRINT "PRESS C TO CONTINUE";: GET Z$: GOTO 1300
```

Optional Computer
Extension: Ratio

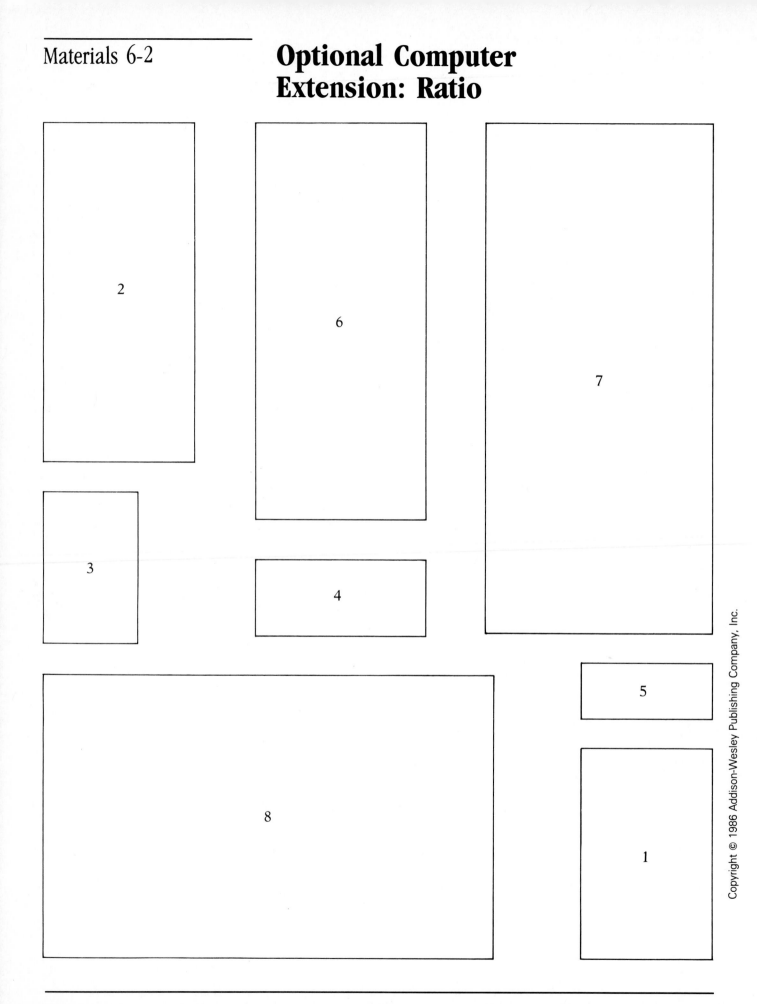

Optional Computer
Extension: Nesting

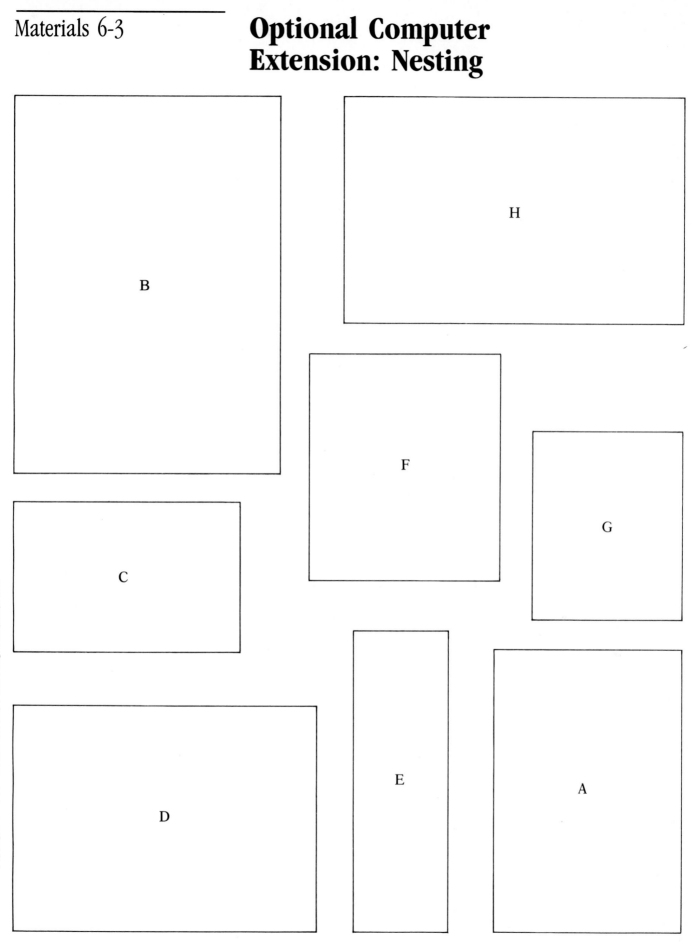

Activity 7

POINT PROJECTIONS

OVERVIEW

In this activity, students learn a third method of drawing similar figures, which is called point projections. This can be thought of as a more accurate rendition of the VTPD (Stretcher) method. The anchor point is now called the point of projection and rather than using two rubber bands to produce a figure twice as large and three rubber bands for a figure three times as large, etc., a stiff paper strip is used to measure the distance from the projection point P to each point. This distance is then used the appropriate number of times (equal to the scale factor) to find the image point. Using a paper strip rather than a ruler gives better final results and allows students to focus on the essentials of the process and not get sidetracked by the precision of measurement.

By moving the point of projection, students discover that the position of the projection point only affects the location of the image, not the size. The scale factor determines the size and orientation of the image.

If the scale factor is greater than 1, the projected image is an enlargement; the image and projection point lie on opposite sides of the original. If the scale factor is between 0 and 1, the image is a reduction; the image lies between the projection point and the original. If the scale factor is negative, the distance is measured in the opposite direction from P so that P is between the original and the image. If the scale factor is between 0 and -1, the image is a reduction. If the scale factor is less than -1, the image is an enlargement. Positive scale factors determine an image that has the same orientation as the original. Negative scale factors determine images that are rotated through 180° from the original. Below is an illustration of point projections of ABC. A is projected to A' in each case.

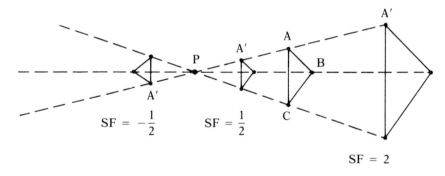

By reptiling the image of the original figures, students discover how the area grows relative to the scale factor. The question of how area grows as the original figure is changed in a specified way has been asked in previous activities. This is a difficult idea that students need to meet several times in many contexts before they are able confidently to explain that area grows by the square of the scale factor. Students should be able to explain that if we start with a figure and enlarge it by a scale factor of S, the new figure will have an area that is S^2 times as large as the original area. Reptiling is a very useful way to help students see this idea more concretely with triangles and rectangles.

Activity 7

Copyright © 1986 Addison-Wesley Publishing Company, Inc.

Goals for students

1. Draw similar figures using point enlargements.
2. Discover that the point of projection only determines the position of the image (new figure).
3. Discover that the scale factor only determines the size of the image.
4. Use reptiling to compare the areas of two similar triangles.

Materials

Several stiff paper strips, about 10 cm by 1.5 cm, for each student.
*Triangles and Projection Points (Materials 7-1).
*Growing Triangles (Materials 7-2).

Worksheets

7-1, Sighting Experiment.
7-2, Find the Projection Point.
7-3, Going Backwards.
7-4, Growing Triangles.
7-5, Moving the Projection Point.
7-6, Shrink Morris (Optional).
7-7, An Invertible (Optional).

Transparencies

Starred items should be made into transparencies.

TEACHER ACTION	TEACHER TALK	EXPECTED RESPONSE
Note: Worksheet 7-1 is optional. If you choose, you may skip it and go directly to Worksheet 7-2, which is launched after Worksheet 7-1.		
Each pair of students should have Reptiles 1, 2, and 3 from Worksheet 4-3, Super Reptiles, and a ruler.	Place Reptile 2 on the desk. Hold Reptile 1 above it in a plane parallel to the desk top and look with one eye at Reptile 1 so that it just hides Reptile 2. Have your partner measure the distance from your eye to Reptile 1 and from your eye to Reptile 2 and record it on Worksheet 7-1.	Eyeball ◉
Have them perform the sighting experiment.		distance A
Demonstrate how to hold the reptiles and how to measure the distances.		distance B
		Reptile 1 ——
Pass out Worksheet 7-1, Sighting Experiment.	Change the distance of Reptile 1 above Reptile 2. Reposition your eye so that 1 hides 2 exactly and measure again. Repeat with a new distance between the reptiles.	Reptile 2 ———
Encourage students to measure accurately, so that the scale factors are seen in the patterns of distances.		Desk top
	Repeat the experiment with Reptiles 1 and 3 and then with 2 and 3. You should have several measures of pairs of distances—A–B, A–C, and B–C—when you finish.	
Ask.	What do you observe about the distances?	$2A = B$
		$3A = C$
		$\frac{3}{2}B = C$
Pass out Worksheet 7-2, Find the Projection Point, and rulers. Ask.	If the large triangle at the top is exactly hidden by the smaller triangle, how can we find the location of the person's eye?	Various answers.

101

TEACHER ACTION	TEACHER TALK	EXPECTED RESPONSE
Demonstrate how to find the point. Have students work along with you.	Connect each pair of corresponding vertices and extend the lines. The lines meet at the location of the person's eye. Another name for the location of the person's eye is the *projection point*, or point of projection.	
	Find the projection point for the bottom pair of triangles.	
Pass out Worksheet 7-3, Going Backwards.	The large shaded triangle could be a shadow cast by any of the four small triangles. Find the projection point in each case.	
Summarize:	How do the small triangles compare to each other?	They are the same size.
	How do the small triangles compare to the large triangle?	The large triangle is four times as large in area as the small triangle.
	Does the distance from the point affect the size of the triangle?	No
	What is the scale factor connecting the smaller triangles to the larger triangle on Worksheet 7-3?	2
	What about on Worksheet 7-2?	2 for the first pair. 3 for the second pair.
Explain.	In Activity 1 we used a VTPD, or Stretcher, to make an enlargement. What does VTPD stand for?	Variable Tension Proportional Divider.
	We are going to learn to make a point projection based on the same principles we used with the VTPD. However, here we will be able to be very precise. If we are careful, the resulting figures for each of us will be the same.	

Activity 7 *Launch*

TEACHER ACTION	TEACHER TALK	EXPECTED RESPONSE
Display a transparency of Materials 7-1, Triangles and Projection Points on the overhead.	Here we have a triangle and a point marked P. The point will be the projection point for the new figure that we will draw.	
Draw lines PA, PB, PC.	We'll draw in lines from P to each of the points of the triangle. We will use a scale factor of 3. This means we'll extend each line so that it's three times as long as it was.	
Extend each line to 3 times its original length. Use a paper strip as a ruler. Mark P A on the strip and slide it along the extended line.	Call the new points A_1, B_1, C_1. Join the points to make a triangle similar to triangle ABC.	

103

TEACHER ACTION	TEACHER TALK	EXPECTED RESPONSE
Pass out Worksheet 7-4, Growing Triangles, and stiff paper strips.	Look at Worksheet 7-4. We'll do number 1 together. Use the stiff paper strips to draw with.	
Display Worksheet 7-4 on the overhead.	What's the scale factor for problem 1?	2
	This means that we will make each line from P to a vertex that is twice as long as its original length.	
	Label the corners of the triangle. Draw PA and extend it until it is twice as long. Mark it A_1.	
	Let's find B, without drawing lines.	
	Mark a point P on your strip. Set it on an imaginary line from P to B, and mark where B occurs.	
	Now slide the strip along the imaginary line, until the P on the strip is at the B of the triangle. The B on the strip is at B_1. Mark it.	
Find C_1, mark it, and draw the new triangle.	Now do the same for C. Find C_1, mark it, then draw the triangle.	

Activity 7 *Launch*

TEACHER ACTION	TEACHER TALK	EXPECTED RESPONSE
Display both enlargements.	Look at your new triangle.	
	How does the side A_1B_1 compare with AB?	It is twice as long.
Mark the new triangle.	And A_1C_1 with AC?	They're all twice as large.
	And C_1B_1 with CB?	

TEACHER ACTION	TEACHER TALK	EXPECTED RESPONSE
Use the paper strip to show that all are 3 times as long.	On the enlargement with a scale factor of 3, how does A_1B_1 compare with AB?	It is 3 times as large.
	How does A_1C_1 compare with AC?	It is 3 times as large.
	And C_1B_1 with CB?	It is 3 times as large.
	What if P were at the top left corner? What do you think we'd get?	Various answers.
	One of the activities is an investigation of this question.	

OBSERVATIONS

POSSIBLE RESPONSES

Students complete Worksheet 7-4, Growing Triangles.

Pass out Worksheet 7-5, Moving The Point.

As students move into this activity ask them to predict what effect moving the projection point will have on the final figures. Many students will say that the farther away the center is, the larger the resulting figure will be.

If students have difficulty, suggest that they draw faint dotted lines from P through the vertex of the triangle, then mark off the desired distance on the line.

Activity 7 *Summarize*

TEACHER ACTION	TEACHER TALK	EXPECTED RESPONSE
Ask.	Let's look at how the area grows as we enlarge the triangles on Worksheet 7-4. The easiest way to see this is to mark the enlargements as a reptile. This works because we're dealing with triangles.	
Display a transparency of Materials 7-2 on the overhead.	In number 1 the scale factor was 2. Each of the sides of the larger triangle is how many times as large as the original?	It is 2 times as large.
Mark mid-points of the sides and then draw in the dotted line.	We can show this by marking the mid-point of each side of the larger triangle.	
	Now we draw in lines to connect these mid-points like a reptile.	
	How many of the smaller triangles does it take to make the larger triangle?	4
	The area of the larger triangle is how many times as great as the area of the smaller one?	It is 4 times as large.
	Now, let's look at number 2 and reptile the large figure.	
	First we need to mark off the sides into thirds because the scale factor was 3.	
	Then we draw in the lines to show a reptile. Notice that every line we drew connected the points we marked as thirds of the sides, *and* every line drawn is parallel to a side of the large triangle.	
	Now how many small triangles does it take to cover the large triangle?	9 of the small triangles.

107

Activity 7 *Summarize*

TEACHER ACTION	TEACHER TALK	EXPECTED RESPONSE
As you go over this, emphasize that the new sides are 3 times as long in number 2 and 4 times as long in number 3. You can mark the length of a side of the original triangle on your paper strip and lay this off on the larger triangle as many times as it fits to get the divisions you need.	So the area of the enlargement is 9 times as great as the original area.	
	What do you predict that we will find when we reptile the enlargement from a scale factor of 4?	Various answers. Some may say 16 small triangles.
	Let's mark it off and see. This time we must divide each side into fourths.	
	This gives us 16 triangles.	
Organize data on the overhead.	So far we have this information.	
Scale Factor Area increases by		
2 4, or 2 × 2		
3 9, or 3 × 3		
4 16, or 4 × 4		
	If we used a scale factor of 5, by how much would the area increase?	5^2 or $5 \times 5 = 25$ times as great.
	How many of the original triangles would cover the enlargement?	25.

Activity 7 *Summarize*

TEACHER ACTION	TEACHER TALK	EXPECTED RESPONSE
	Now, let's look at Worksheet 7-5.	
	Do the enlargements get bigger when the point is farther away?	No.
	What does moving the point do to the enlargement?	Moves it around.
	Can it turn the figure upside down?	No.
	If we want an enlargement to be right side up, what must be true?	Start right side up.
Ask. Demonstrate on the overhead.	Look at the second projection on Worksheet 7-4 with a scale factor of 3.	
	Suppose we made a number line with 0 at *P* and 1 at *A*. What other number could we assign?	
Ask.	What do these numbers have to do with the projection?	They tell the scale factor.
Ask.	Where would $\frac{1}{2}$ be on the number line?	Between 0 and 1.
Ask.	What would a projection with a scale factor of $\frac{1}{2}$ do to the triangle?	Shrink it.

TEACHER ACTION	TEACHER TALK	EXPECTED RESPONSE
Point to the left of *P*.	What numbers would be over here?	Negatives.
	What would a scale factor of −1 do?	Keep the same size and change the orientation.
	An example of a negative projection is the eye of a camera or a human eye. Both lenses turn the image upside down and shrink or enlarge it depending on the scale factor.	
Assign Worksheet 7-6, Shrink Morris, and Worksheet 7-7, An Invertible, as an extra challenge or as homework.	Worksheet 7-6 and 7-7 are examples of a fraction as a scale factor and a negative scale factor.	
	Remember that a scale factor of one half will shrink the projected image so that Morris's image will be between the point of projection and Morris.	
	With a scale factor of −1, the image of the car will be on the opposite side of the projection point from the original.	

Triangles and Projection Points

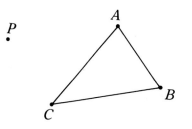

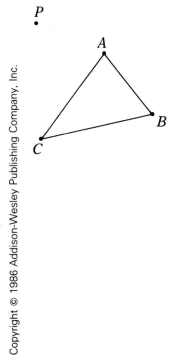

Growing Triangles

1. Scale Factor 2

2. Scale Factor 3

3. Scale Factor 4

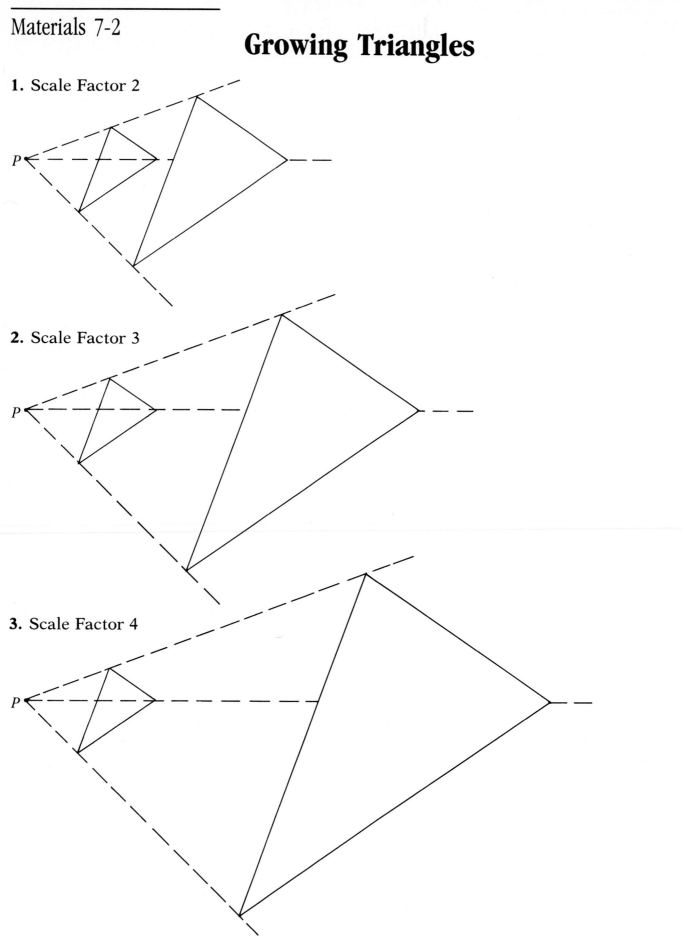

Sighting Experiment

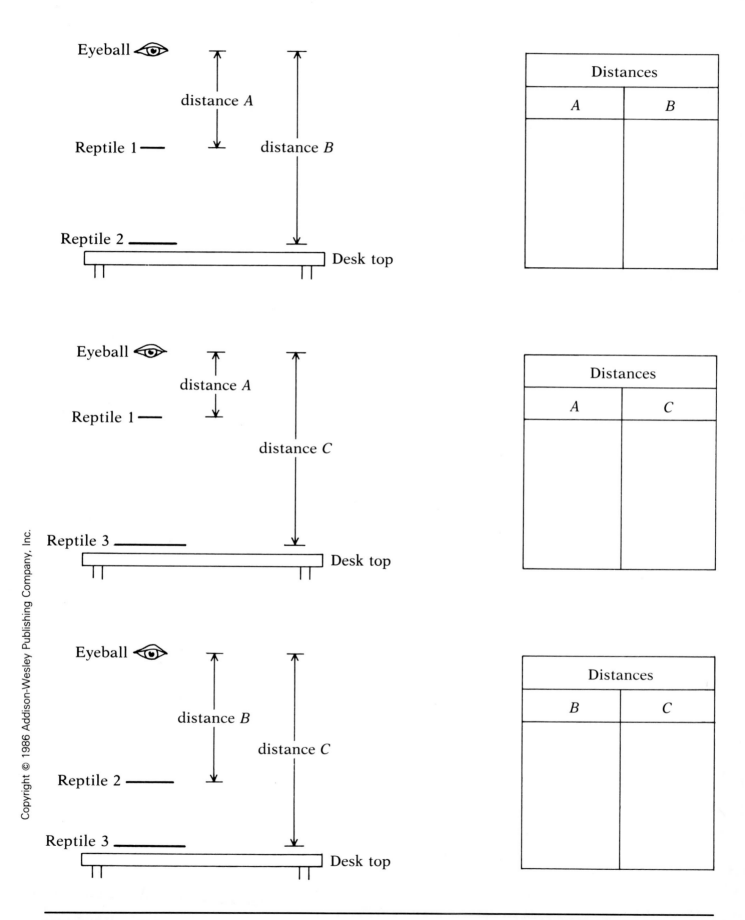

Distances	
A	*B*

Distances	
A	*C*

Distances	
B	*C*

Worksheet 7-1

Find the Projection Point

1. For each pair of triangles, find the projection point that projects the smaller triangle to the larger.

2.

Going Backwards

Show how the large triangle is a point projection from each of the smaller triangles. Label the projection points.

What is the scale factor from the smaller triangle to the larger? _____

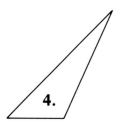

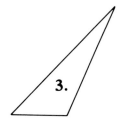

Growing Triangles

Use *P* as the projection point and the given scale factor to find a new triangle. In each problem, show how the dimensions of the new triangle compare to the original.

1. Scale Factor 2

P •
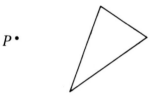

2. Scale Factor 3

P •

3. Scale Factor 4

P •

Worksheet 7-4

Moving the Projection Point

Use a scale factor of 2 and draw the resulting picture for each of the centers P_1, P_2, P_3 and P_4.

P_1

P_2

P_3

P_4

Shrink Morris

Use scale factor $\frac{1}{2}$ and P as the projection point to shrink Morris.

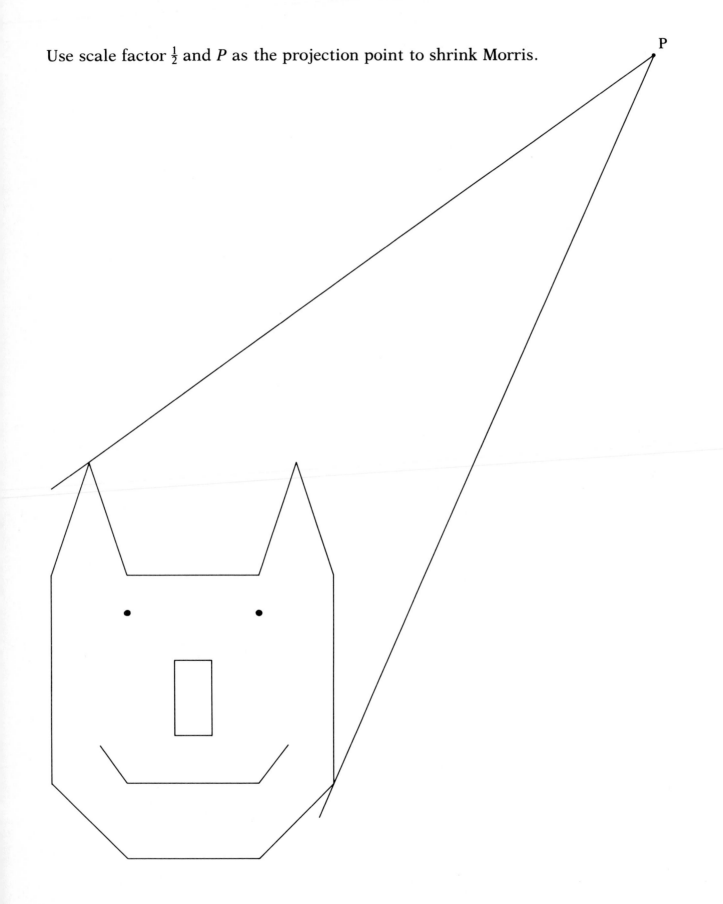

P

Worksheet 7-6

An Invertible

Use a scale factor of -1 and P as a projection point to flip the car.

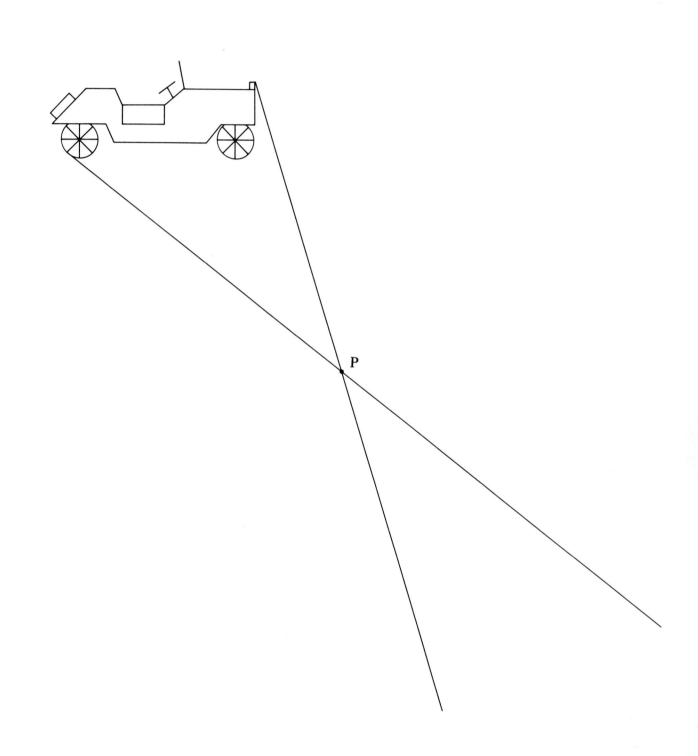

Activity 8

PICTURE ENLARGEMENTS

OVERVIEW

This activity is the capstone for the unit theme of exploring how area grows as lengths change in similar figures. Students first enlarge a picture cut from a magazine. This is an extension of Activity 7; this time, however, the students must position all three parts: original, image, and projection point. The following guidelines may be helpful:

1. The initial enlargement should have a scale factor of 2.

2. The projection point must be far enough away so that the image falls clear of the original, yet close enough that the angle is as large as feasible.

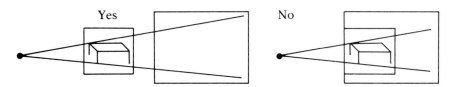

3. Let students bring in cartoons, pictures, or drawings to enlarge. Simple pictures with mostly straight lines work best. Drawings work better than photographs, which are not so precise.

4. If the original has curved lines, one strategy is to mark dots, find the projection of the dots, and connect them.

5. Use tape if necessary to keep the projection point, original, and image in fixed positions.

After completing the enlargements, students perform a series of measurements to compare the growth of lengths and areas in similar figures.

Finally, a series of scale-model application problems allows students to use their knowledge to solve real problems.

Goals for students

1. Decide where to place picture and projection point.
2. Practice making point enlargements.
3. Use the fact that the area growth factor is the square of the scale factor to solve application problems.

Materials

Pictures from magazines.
Stiff paper strips.

Worksheets

*8-1, Bigger Tables.
 8-2, Application Problems.

Transparencies

Starred item should be made into a transparency.

121

TEACHER ACTION	TEACHER TALK	EXPECTED RESPONSE
Students need Worksheet 8-1, Bigger Tables, and a simple cartoon or picture consisting of mostly straight lines. Each student needs a supply of paper strips to measure length.	Today you get to enlarge the pictures that you have found. To be sure that you understand what you need to do, let's look at an example together. You should mark your figure on Worksheet 8-1 as I do it at the overhead.	
Display a transparency of Worksheet 8-1.	Our task is to enlarge the table using a scale factor of 2. We select the points that would help us draw the enlargement. We need the four corners of the table and the ends of the legs.	
Mark the corners and leg ends to emphasize the needed points.	Now we proceed as we did in Activity 7 and locate the points for the new table by doubling the distance from *P*.	
Carefully illustrate finding the points and finishing the enlargement.		
	Now let's answer question (b), "What is the length of the small table?"	3 cm
	What should be the length of the large table?	$2 \times 3 = 6$ cm
You can reptile the large table top to show the 4 smaller tables.	How should the areas compare?	The enlargement has an area 4 times as large.

OBSERVATIONS

POSSIBLE RESPONSES

Tape your cartoon or picture to your desk so that it doesn't slide. Tape a clean piece of paper beside the cartoon. Choose a projection point carefully so that your figure will land on the paper. Experiment with a few points to test the location of P before you proceed.

Some students will need help locating a suitable projection point.

As students near completion of their drawings, remind them to compare some length and some areas in the two drawings.

TEACHER ACTION

TEACHER TALK

EXPECTED RESPONSE

Give several students a chance to show their pictures and explain what lengths and areas they compared.

Be sure to restate that lengths grow by the scale factor and area by the *square* of the scale factor.

TEACHER ACTION	TEACHER TALK	EXPECTED RESPONSE
Students need Worksheet 8-2, Application Problems.	Let's see if we can use what we have learned to answer some problems about scale models.	
Read the first part of number 1.	It costs Mrs. Jones $400 to carpet the small room shown. She has another room twice as long and twice as wide. How much will it cost to carpet the bigger room in the same carpeting?	
Illustrate on overhead.		Some students will answer $800. Respond by saying "Let's check this out by drawing the bigger room."
$400		
Illustrate.	Draw a picture of the bigger room.	
	Are we concerned with area or length when we buy carpet?	Area; carpet *covers* the floor of room.
	Good. How many small room floors fit into the big room floor?	4
Show the 4 small rooms.	So how much will it cost to carpet the room?	$4 \times \$400 = \$1600.$
If you feel the class needs it, do the second part before you assign the rest for class or homework.	As you work these problems, be sure to draw a picture of each situation to help you decide whether you are concerned with length or area growth.	
Application problems are hard for some students, so you will probably need to discuss the answers to these questions carefully after the students have attempted the assignment.		

Bigger Tables

1. a. Use a scale factor of 2 and the projection point P to enlarge the table.

$P \bullet$

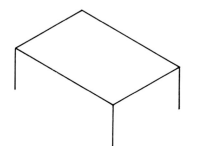

b. What is the length of the small table? _____

Use the scale factor to predict the length of the big table. _____
Measure to check your prediction.

c. Use the scale factor to predict how the areas compare. _____

_____

_____

Find the area of each table top to check your prediction.

_____

2. Make an enlargement of your cartoon or picture, using a scale factor of 2. Measure some length on your original and on the enlargement. If you are careful the length in the big picture should be twice the length in the original.

Find the area of some part of your picture and the enlargement. Is the enlargement four times as big in area?

Application Problems

In each problem, first draw a picture. Then decide whether the problem involves measuring length or area growth.

1. It costs Mrs. Jones $400 to carpet the small room shown. She has another room twice as long and twice as wide. How much will it cost her to carpet the bigger room in the same carpeting? Before you answer, draw a picture of the bigger room. Is the question asked about the growth of length or of area?   _____

Suppose Mrs. Jones wanted to put the same carpet in a room 3 times as long and 3 times as wide. How much would it cost?   _____

2. Sue is building a model boat. The scale factor from the model boat to the real boat is 20. If the big boat is 40 feet long, how long is the model?   _____

If the model boat has doors that are 2 inches high, how high are the doors on the real ship.   _____

If it takes $\frac{1}{3}$ can of paint to paint the deck of the model boat, how much paint will it take to paint the deck of the real boat?   _____

3. The toy shop has a doll house copy of a real house built on a scale of 1' to 10'. This means that the scale factor from the doll house to the real house is 10. If the picket fence around the doll house is 14 feet long, how long is the picket fence around the real house?   _____

If it takes 200 yards of fabric to make the curtains and bedspreads for the real house, how much fabric is needed for the doll house?   _____

The toy house has 3 chimneys, how many chimneys does the real house have?   _____

Worksheet 8-2

Activity 9

APPLICATIONS USING SIMILAR TRIANGLES

Copyright © 1986 Addison-Wesley Publishing Company, Inc.

OVERVIEW

The height of an inaccessible object can be found by using similar triangles. The similar triangles used are those formed by shadows or by angles of reflections in a mirror. The shadow method can be used only on a reasonably sunny day.

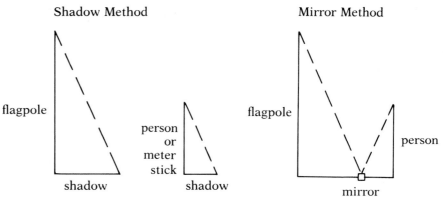

Goals for students

1. Recognize that the ratio of the lengths of the shadows cast by two objects is equal to the ratio of the heights of the two objects.
2. Recognize that the angle of reflection in a mirror creates two similar triangles.
3. Find the height of a flagpole (or some other inaccessible object) by using the lengths of shadows.
4. Find the height of a flagpole by using a mirror.

Materials

Two meter sticks for each group of students.
One small mirror for each group of students.

Worksheets

*9-1, Measuring with the Shadow Method.
*9-2, Measuring with the Mirror Method.
*9-3, Indirect Measurement.

Transparencies

Starred items should be made into transparencies.

APPLICATIONS USING SIMILAR TRIANGLES

TEACHER ACTION	TEACHER TALK	EXPECTED RESPONSE
Ask.	There is a flagpole in front of a school building. How tall do you think it is?	Record guesses from the class.
Students need Worksheet 9-1, Measuring with the Shadow Method, and Worksheet 9-2, Measuring with the Mirror Method.	What are some ways that we could find out how tall it is?	Rule out climbing.
	We can use the similarity ideas that we have been studying to find out the height.	
Illustrate using a transparency of Worksheet 9-1.	If I were out by the flagpole on a sunny day and held a meter stick perpendicular to the level ground, what would happen?	It would cast a shadow.
	Would the flagpole also cast a shadow?	Yes.
	Are there similar triangles involved? Illustrate.	Yes.
	How do we know that they are similar?	Both are right triangles, and corresponding angles are equal; if you line up the right angles, then the hypotenuses are parallel.
Be sure students record the measures and computation for the example on their worksheets.	Let's study the similar triangles involved to see what we can measure. Assume the following measures were obtained. meter stick = 1 m meter stick shadow = .2 m flagpole shadow = 6 m	Measure the meter stick, the meter stick shadow, and the flagpole shadow.
Work through the example.	Record these on your example space on Worksheet 9-1. Here is one way to use the similar triangles.	

TEACHER ACTION

Illustrate.

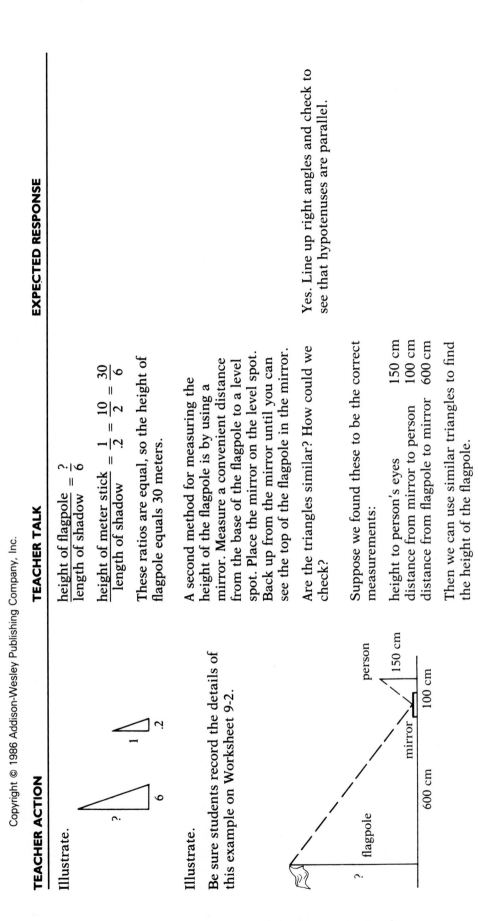

Illustrate.

Be sure students record the details of this example on Worksheet 9-2.

TEACHER TALK

$$\frac{\text{height of flagpole}}{\text{length of shadow}} = \frac{?}{6}$$

$$\frac{\text{height of meter stick}}{\text{length of shadow}} = \frac{1}{.2} = \frac{10}{2} = \frac{30}{6}$$

These ratios are equal, so the height of flagpole equals 30 meters.

A second method for measuring the height of the flagpole is by using a mirror. Measure a convenient distance from the base of the flagpole to a level spot. Place the mirror on the level spot. Back up from the mirror until you can see the top of the flagpole in the mirror.

Are the triangles similar? How could we check?

Suppose we found these to be the correct measurements:

height to person's eyes   150 cm
distance from mirror to person   100 cm
distance from flagpole to mirror 600 cm

Then we can use similar triangles to find the height of the flagpole.

EXPECTED RESPONSE

Yes. Line up right angles and check to see that hypotenuses are parallel.

Measuring with the Shadow Method

Problem

To find how tall the flagpole is.

Method

Measure the length of the shadow of a meter stick and the length of the shadow of the flagpole. Use centimeters. Record your measurements in the appropriate place on the picture.

Computation

Form a ratio using similar triangles and solve the problem.

Example

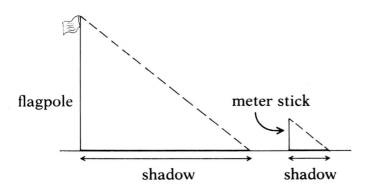

flagpole   meter stick

shadow   shadow

Real measurements

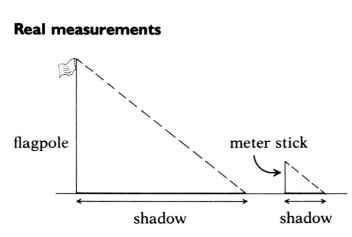

flagpole   meter stick

shadow   shadow

Measuring with the Mirror Method

Problem

To find how tall the flagpole is.

Method

Measure a *convenient* distance from the bottom of the flagpole to a level spot of ground. Make this distance something like 300 to 500 cm from the base.

Place the mirror at the spot measured. Back up until you can see the top of the flagpole in the mirror.

Measure the distance from the mirror to your feet.

Measure the height from the level ground to your eyes.

Record these measurements on the picture.

Example

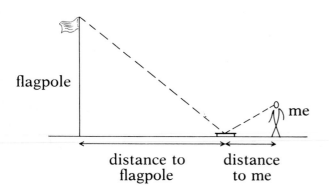

flagpole

me

distance to flagpole   distance to me

Real measurements

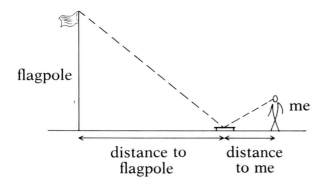

flagpole

me

distance to flagpole   distance to me

Worksheet 9-2

NAME _____

Indirect Measurement

1. The picture below shows how Mary used a short tree to find the height of the tall tree.

   What triangles are similar? _____

   What is the height of the tall tree?

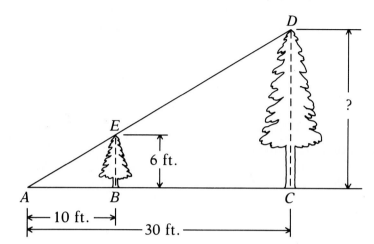

2. José used a stick and shadows to find the height of a tree. What answer should he get?

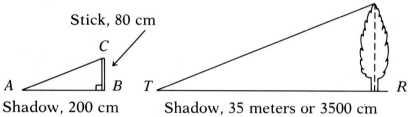

Stick, 80 cm

Shadow, 200 cm   Shadow, 35 meters or 3500 cm

3. Paul and Samantha want to find how wide a natural rock tower is. First they locate a point *A* from which they can sight the edge of the rock on both sides. From *A* they draw lines to both edges of the rock and beyond. Then they stake points *B* and *C* so that the line from *B* to *C* is parallel to the width from *D* to *E* that they want to measure. Then they measure the distances, as shown in the diagram. How wide is the rock tower?

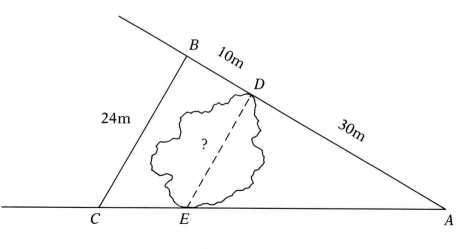

Indirect Measurement

4. Find the distance across the pond.

What triangles are similar? _____

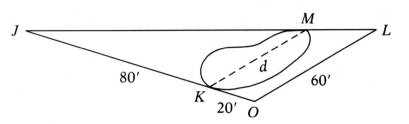

5. Stand 9 ft from a friend and hold a 1-foot ruler in front of yourself. Line up the top of the ruler with the top of your friend and the bottom of the ruler with the feet of your friend. If the ruler is 2 ft from your eyes, how tall is your friend?

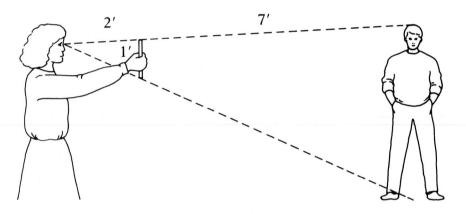

6. Sue used a mirror to find the height of a tree. Her eyes are 5 ft 6 in. from the ground and she finds $\overline{GM}$ = 3 ft and $\overline{MD}$ = 21 ft. How high is the tree?

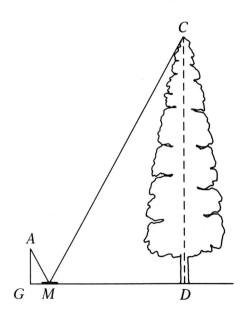

Optional Computer Extension
For Indirect Measurement

The following program will perform all computations necessary to use the shadow and mirror methods of indirect measurement. As listed, it will run on Apple computers. Directions for adapting the program to other machines are given below.

The user first specifies the object to be measured, units used, and method of measurement. (The underlined words are the user's responses.)

```
INDIRECT MEASUREMENT
WHAT ARE YOU MEASURING? TREE
WHAT UNITS? FEET
WHICH METHOD?
SHADOWS OR MIRRORS
PRESS S OR M
```

The screen then displays a diagram and the user enters his/her measurements:

```
HEIGHT OF STICK? 2
SHADOW OF STICK? 1.5
SHADOW OF TREE? 8
HEIGHT OF TREE
  10.7 METERS
PRESS C TO CONTINUE
```

```
EYE LEVEL OF VIEWER? 1.8
VIEWER TO MIRROR? 3
MIRROR TO TREE? 5.9
HEIGHT OF TREE
  3.5 METERS
PRESS C TO CONTINUE
```

Indirect Measurement

```
50   DIM A$(10),B$(10),C$(10)
100  A$(1) = "EYE LEVEL OF VIEWER  "
110  A$(2) = "HEIGHT OF STICK "
120  B$(1) = "VIEWER TO MIRROR "
130  B$(2) = "SHADOW OF STICK  "
140  C$(1) = "MIRROR TO "
150  C$(2) = "SHADOW OF "
400   HOME : PRINT "INDIRECT MEASUREMENT"
405   PRINT
410   PRINT : PRINT "NAME OF OBJECT BEING MEASURED"
420   INPUT O$
430   PRINT : PRINT "WHAT UNITS ";: INPUT U$
440   PRINT : PRINT
500   PRINT "WHICH METHOD?"
510   INVERSE : PRINT "SHADOWS";: NORMAL : PRINT " OR ";: INVERSE : PRINT
     "MIRRORS": NORMAL
520   PRINT : PRINT "PRESS ";: INVERSE : PRINT "S";: NORMAL : PRINT " OR
     ";: INVERSE : PRINT "M": NORMAL
530   GET RP$: IF RP$ < > "S" AND RP$ < > "M" THEN 530
540   IF RP$ = "S" THEN M = 2: GOTO 1000
550   IF RP$ = "M" THEN M = 1: GOTO 2000
1000   HOME
1010   VTAB (10): INVERSE : FOR J = 1 TO 7: PRINT " ": NEXT J
1015   NORMAL
1020   VTAB (14): FOR J = 1 TO 3: HTAB (13): PRINT "|": NEXT J
1025   VTAB (16): HTAB (1)
1030   PRINT "=======";: HTAB (13): PRINT "==="
1040   FOR J = 1 TO 30: PRINT "-";: NEXT J: PRINT
1050   PRINT
1100   PRINT A$(2);: INPUT A
1110   PRINT B$(2);: INPUT B
1130   PRINT C$(2);O$;: INPUT C
1140   PRINT : PRINT "HEIGHT OF "O$
1150   PRINT  INT (A * C / B * 10 + .5) / 10;" "U$
1200   PRINT : PRINT "PRESS C TO CONTINUE"
1210   GET RP$: IF RP$ < > "C" THEN 1210
1220   GOTO 400
2000   HOME
2010   VTAB (7)
2020   FOR J = 1 TO 3: INVERSE : PRINT " ";: NORMAL : PRINT  SPC( J - 1);
     "\"
2025   NEXT J:
2030   INVERSE : PRINT " ";: NORMAL : PRINT  SPC( 3);"\" SPC ( 7);"0"
2040   FOR J = 1 TO 3: INVERSE : PRINT " ";: NORMAL : PRINT  SPC( J + 3);
     "\"; SPC( 7 - 2 * J);"/"; SPC( J - 1);"|"
2045   NEXT J
2050   INVERSE : PRINT " ";: NORMAL : PRINT  SPC( 6);"---"; SPC( 2);"|"
2060   FOR J = 1 TO 20: PRINT "-";: NEXT J
2065   PRINT
2070   PRINT A$(1);: INPUT EL
2080   PRINT C$(1);O$;: INPUT OM
2090   PRINT B$(1);: INPUT MF
3000   PRINT "HEIGHT OF "O$" IS"
3010  HT = OM * EL / MF:HT =  INT (HT * 10 + .5) / 10
3020   PRINT HT" "U$
3030   GOTO 1200
```

Review Problems

1. The rectangles A, B, C, D have been piled so that their lower left corners are at 0.

Which two rectangles are similar? _____

A. C and D only

B. B and D

C. A and B

D. A and C

E. A and D

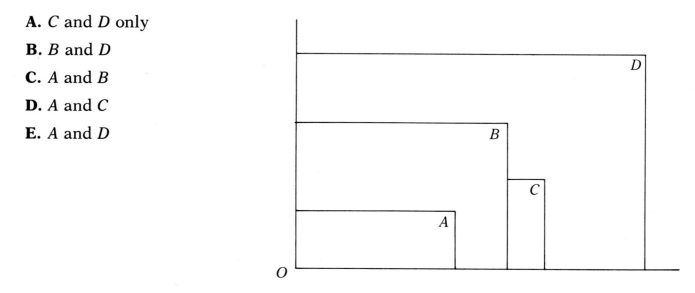

2. Triangle ABC is enlarged by a point projection from P.

What is the scale factor? _____

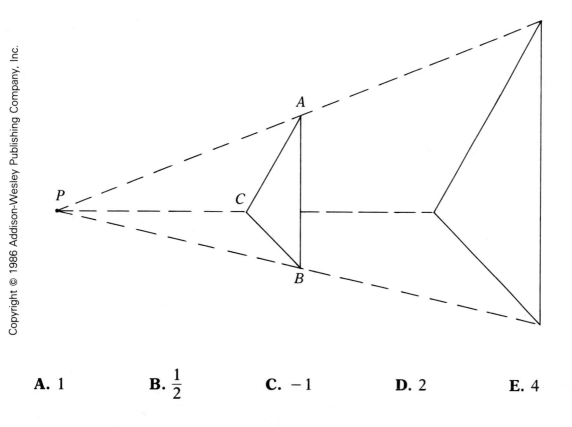

A. 1 **B.** $\frac{1}{2}$ **C.** -1 **D.** 2 **E.** 4

Review Problems

3. The two triangles are similar. Find the missing measurements.

x = _____   y = _____

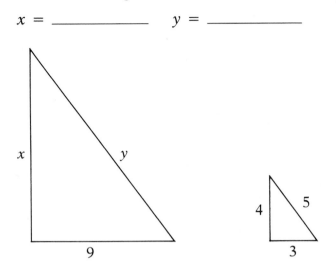

4. The two triangles are similar. Find the missing measurement. _____

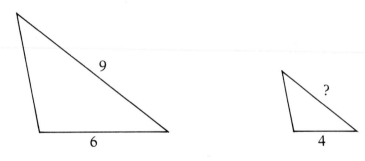

5. All three rectangles are similar. Find the missing measurements.

Rectangle B _____   Rectangle C _____

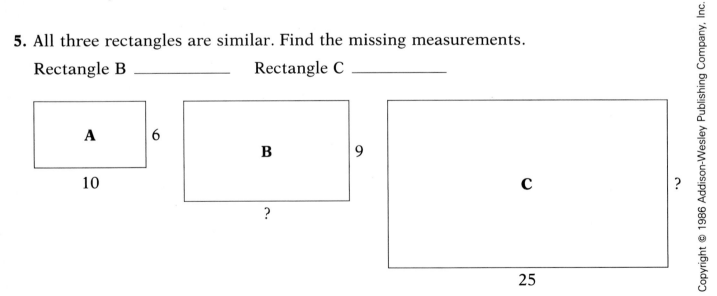

Review Problems

6. Use the measurements of the shadows and the rod to find the height of the tree.

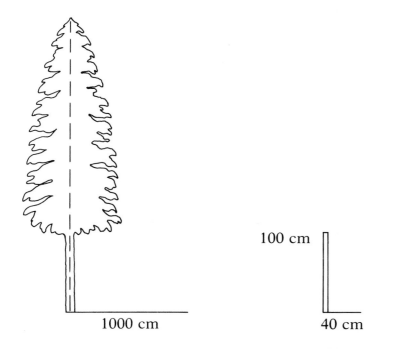

100 cm

1000 cm   40 cm

7. The dimensions of 6 rectangles are given below. List the groups of similar rectangles.

Rectangle	Short Side	Long Side
A	18	21
B	15	24
C	24	28
D	$2\frac{1}{2}$	4
E	12	14
F	3	$3\frac{1}{2}$

8. Divide the triangle into 4 congruent reptiles.

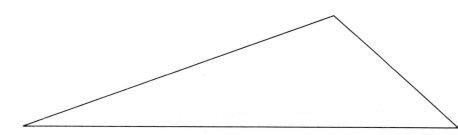

Review Problems

9. Use the line segment *AB* as a base, and draw a rectangle that is similar to the small rectangle.

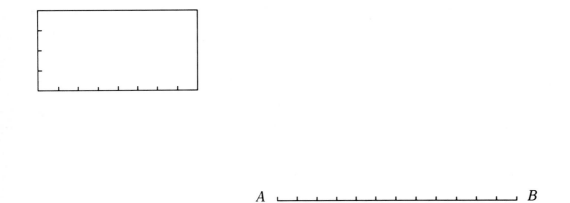

A └─┴─┴─┴─┴─┴─┴─┴─┴─┴─┘ B

10. Construct an arrowhead enlarged by a scale factor of 3. (The guidelines have been drawn in to help you.)

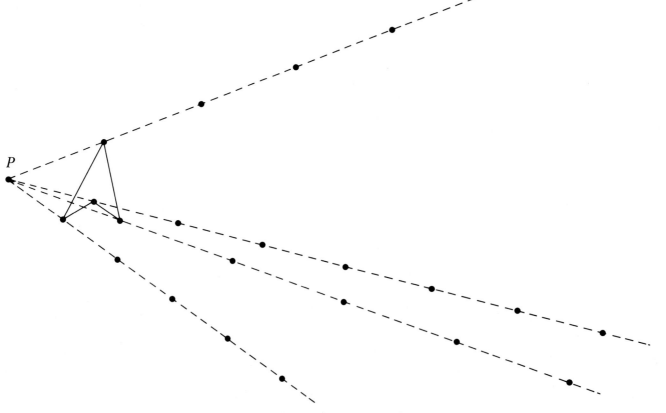

Review Problems, page 4

Review Problems

11. The original rectangle A has been enlarged using several different scale factors. Complete the table.

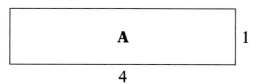

Rectangle	Scale Factor	Short Side	Long Side	Perimeter	Area
A	1	1	4		
B	2				
C	3				
D	$\frac{1}{2}$				
E	$\frac{3}{2}$				

12. Construct an arrowhead enlarged by a scale factor of 4. (The guidelines have been drawn in to help you.)

Unit Test

1. Given rectangle:

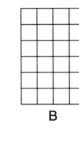

Which of the following rectangles is similar to the given rectangle?

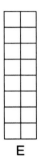

A   B   C   D   E

2. A 1 × 5 rectangle grows into a 4 × 20 rectangle. The area of the new rectangle is how many times larger than the area of the small rectangle?

A 3 times   B 4 times   C 5 times   D 15 times   E 16 times

3. The given figures are similar. Find the missing length.

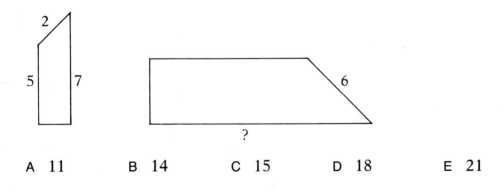

A 11   B 14   C 15   D 18   E 21

4. The given rectangles are similar. Find the length of the missing side.

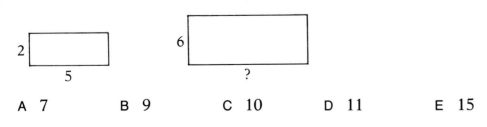

A 7   B 9   C 10   D 11   E 15

Unit Test, page 1

Unit Test

5. A picture of a tall building is taken. The following measurements are known.

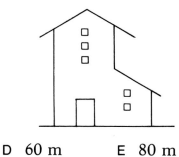

	Picture	Actual
height of door	10 mm	2 m
height of building	40 mm	?

What is the actual height of the building?

A 8 m   B 6 m   C 10 m   D 60 m   E 80 m

6. If the lengths of the sides of a triangle are each multiplied by 3, how much larger will the area of the new triangle be?

A   3 times larger
B   6 times larger
C   9 times larger
D   12 times larger
E   15 times larger

7. A given square and its image:

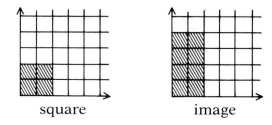

square   image

Which of these transformations was used?

A $(x, y) \to (x, 2y)$
B $(x, y) \to (2x, y)$
C $(x, y) \to (2x, 2y)$
D $(x, y) \to (2x, 4y)$
E $(x, y) \to (4x, 2y)$

8. The given triangles are similar. Find the missing length.

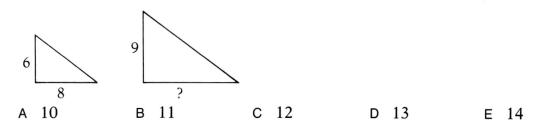

A   10   B   11   C   12   D   13   E   14

Unit Test

9. The two triangles below are similar and the lengths of the sides of the larger triangle are 3 times that of the smaller triangle.

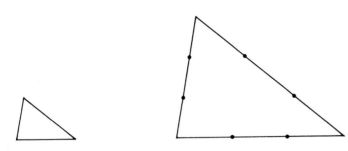

How many of the smaller triangles will exactly fit into the larger one?

A 4   B 6   C 7   D 8   E 9

10. A man who is 6 feet tall has a shadow that is 8 feet long. At the same time, a nearby tree has a shadow that is 32 feet long. How tall is the tree?

A 30 feet   B 21 feet   C 24 feet   D 42 feet   E 48 feet

11. Given rectangles of dimensions 1 × 6 and 4 × 24.

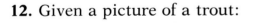
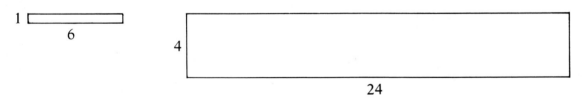

The area of the larger rectangle is how many times bigger than the area of the smaller rectangle?

A 4 times   B 6 times   C 8 times   D 16 times   E 18 times

12. Given a picture of a trout:

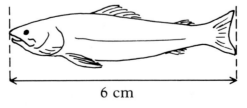

6 cm

The scale of the picture to the real trout is 1 to 12.

What is the actual length of the real trout?

A $\frac{1}{2}$ cm   B 72 cm   C 2 cm   D 144 cm   E 18 cm

Unit Test, page 3

13. Which of the following rectangles is similar to a 10 × 15 rectangle?

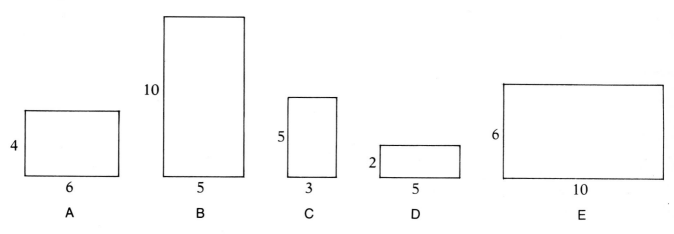

14. Which of these figures is divided into small shapes that are similar to the big shape?

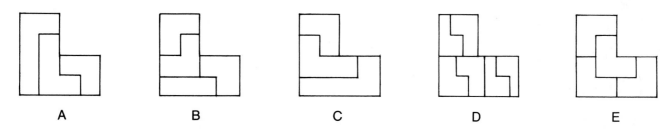

15. The ratio of the sides of a rectangle is $\frac{2}{3}$.

Which of the following could be the ratio of the sides of a similar rectangle?

A $\frac{4}{9}$ B $\frac{4}{3}$ C $\frac{2}{6}$ D $\frac{4}{5}$ E $\frac{6}{9}$

16. John cut out four paper rectangles (W, X, Y, Z) and lined them up as shown below.

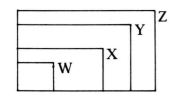

Which pair of rectangles is similar?

A W and X   B X and Y   C Y and Z   D W and Y   E X and Y

Unit Test

17. Which of these figures shows the big triangle divided into smaller triangles that are similar to the big triangle?

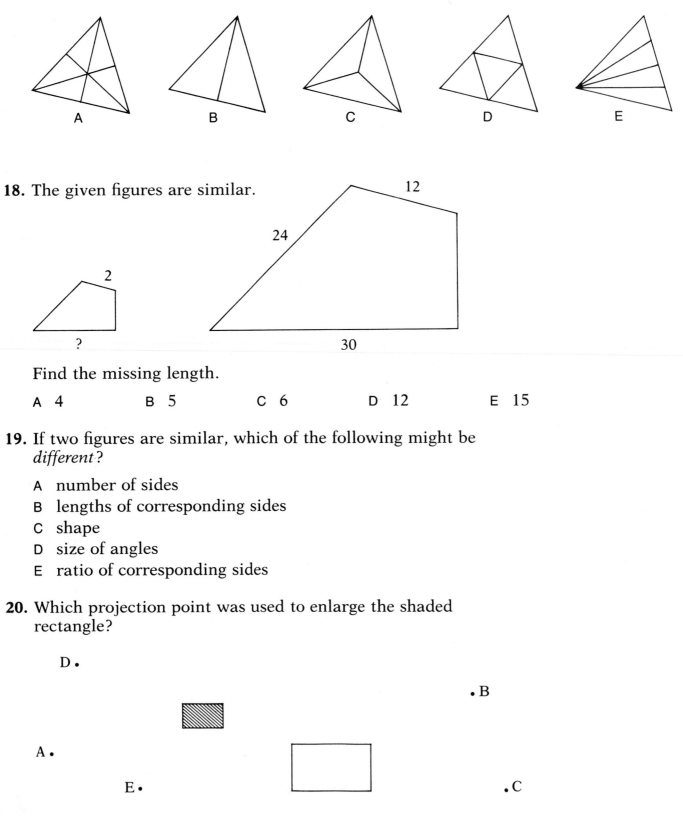

A   B   C   D   E

18. The given figures are similar.

Find the missing length.

A 4   B 5   C 6   D 12   E 15

19. If two figures are similar, which of the following might be *different*?

A number of sides
B lengths of corresponding sides
C shape
D size of angles
E ratio of corresponding sides

20. Which projection point was used to enlarge the shaded rectangle?

D•

•B

A•

E•

•C

NAME

Unit Test

21. Joan estimates the height of a flagpole by using a mirror.

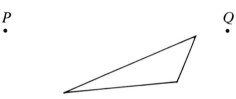

Distances	
to eye level	5 ft
Joan to mirror	2 ft
mirror to pole	10 ft

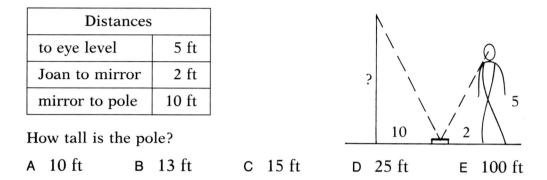

How tall is the pole?

A 10 ft   B 13 ft   C 15 ft   D 25 ft   E 100 ft

22. Point projections are made of the given triangle first from *P* and then from *Q*. Each projection uses a scale factor of 3.

P
•

Q
•

Which of the following is true?

A The triangle image from *P* has larger angles.
B The triangle image from *Q* has larger angles.
C The triangle image from *P* has a larger area.
D The triangle images from *P* and *Q* have the same area.
E The triangle image from *Q* has a larger area.

23. A 2-meter stick has a shadow of $\frac{1}{2}$ m at the same time that a nearby tree has a shadow of 3 m. How tall is the tree?

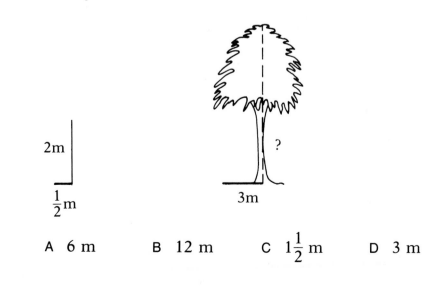

2m

$\frac{1}{2}$ m

3m

?

A 6 m   B 12 m   C $1\frac{1}{2}$ m   D 3 m   E 15 m

24. A given triangle and its image:

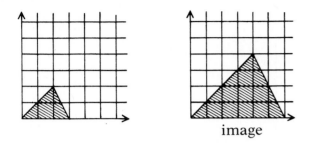

image

Which of these transformations was used?

A $(x, y) \rightarrow (2x, 2y)$
B $(x, y) \rightarrow (x, 2y)$
C $(x, y) \rightarrow (2x, y)$
D $(x, y) \rightarrow (2x, 4y)$
E $(x, y) \rightarrow (4x, 2y)$

25. What scale factor has been used to enlarge the small sailboat?

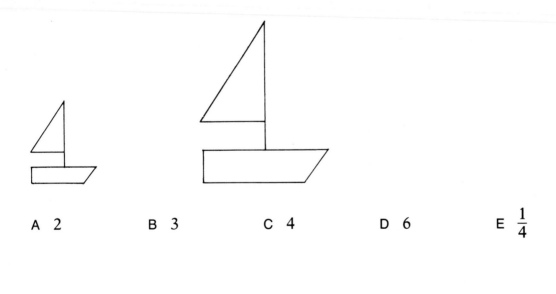

A 2   B 3   C 4   D 6   E $\frac{1}{4}$

Unit Test Answer Sheet

1.	A	B	C	D	E
2.	A	B	C	D	E
3.	A	B	C	D	E
4.	A	B	C	D	E
5.	A	B	C	D	E

6.	A	B	C	D	E
7.	A	B	C	D	E
8.	A	B	C	D	E
9.	A	B	C	D	E
10.	A	B	C	D	E

11.	A	B	C	D	E
12.	A	B	C	D	E
13.	A	B	C	D	E
14.	A	B	C	D	E
15.	A	B	C	D	E

16.	A	B	C	D	E
17.	A	B	C	D	E
18.	A	B	C	D	E
19.	A	B	C	D	E
20.	A	B	C	D	E

21.	A	B	C	D	E
22.	A	B	C	D	E
23.	A	B	C	D	E
24.	A	B	C	D	E
25.	A	B	C	D	E

Answers

NAME _____

Morris

NAME _____

Morris

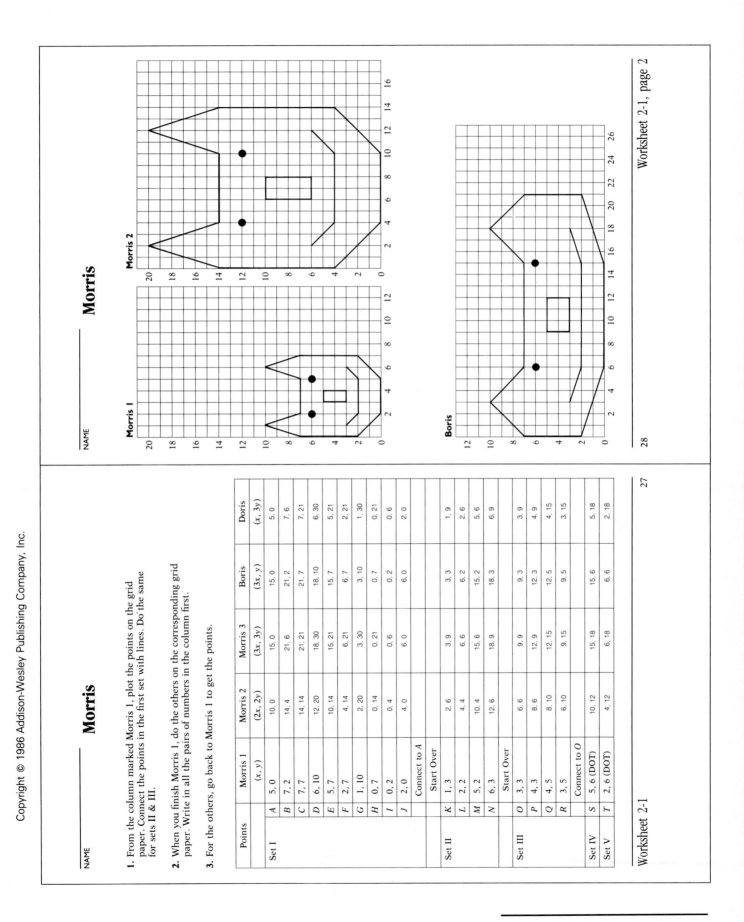

1. From the column marked Morris 1, plot the points on the grid paper. Connect the points in the first set with lines. Do the same for sets II & III.

2. When you finish Morris 1, do the others on the corresponding grid paper. Write in all the pairs of numbers in the column first.

3. For the others, go back to Morris 1 to get the points.

Points		Morris 1 (x, y)	Morris 2 (2x, 2y)	Morris 3 (3x, 3y)	Boris (3x, y)	Doris (x, 3y)
Set I	A	5, 0	10, 0	15, 0	15, 0	5, 0
	B	7, 2	14, 4	21, 6	21, 2	7, 6
	C	7, 7	14, 14	21, 21	21, 7	7, 21
	D	6, 10	12, 20	18, 30	18, 10	6, 30
	E	5, 7	10, 14	15, 21	15, 7	5, 21
	F	2, 7	4, 14	6, 21	6, 7	2, 21
	G	1, 10	2, 20	3, 30	3, 10	1, 30
	H	0, 7	0, 14	0, 21	0, 7	0, 21
	I	0, 2	0, 4	0, 6	0, 2	0, 6
	J	2, 0	4, 0	6, 0	6, 0	2, 0
		Connect to A				
		Start Over				
Set II	K	1, 3	2, 6	3, 9	3, 3	1, 9
	L	2, 2	4, 4	6, 6	6, 2	2, 6
	M	5, 2	10, 4	15, 6	15, 2	5, 6
	N	6, 3	12, 6	18, 9	18, 3	6, 9
		Start Over				
Set III	O	3, 3	6, 6	9, 9	9, 3	3, 9
	P	4, 3	8, 6	12, 9	12, 3	4, 9
	Q	4, 5	8, 10	12, 15	12, 5	4, 15
	R	3, 5	6, 10	9, 15	9, 5	3, 15
		Connect to O				
Set IV	S	5, 6 (DOT)	10, 12	15, 18	15, 6	5, 18
Set V	T	2, 6 (DOT)	4, 12	6, 18	6, 6	2, 18

Worksheet 2-1

Worksheet 2-1, page 2

Answers

Summary of Morris's Noses

NAME

Morris	Rule	Bottom Edge	Side Edge	Ratio: Bottom/Side	Area	Perimeter
1	(x, y)	1	2	$\frac{1}{2}$	2	6
2	$(2x, 2y)$	2	4	$\frac{2}{4}$ or $\frac{1}{2}$	8	12
3	$(3x, 3y)$	3	6	$\frac{3}{6}$ or $\frac{1}{2}$	18	18
4	$(4x, 4y)$	4	8	$\frac{4}{8}$ or $\frac{1}{2}$	32	24
5	$(5x, 5y)$	5	10	$\frac{5}{10}$ or $\frac{1}{2}$	50	30
	$(7x, 7y)$	7	14	$\frac{7}{14}$ or $\frac{1}{2}$	98	42
	$(10x, 10y)$	10	20	$\frac{10}{20}$ or $\frac{1}{2}$	200	60

	Rule	Bottom Edge	Side Edge	Ratio: Bottom/Side	Area	Perimeter
Boris	$(3x, y)$	3	2	$\frac{3}{2}$	6	10
Doris	$(x, 3y)$	1	6	$\frac{1}{6}$	6	14

Do Boris and Doris fit the same patterns as Morris 1, 2, and 3? No

Does Morris 99 fit in the same patterns as Morris 1, 2, and 3? Yes

Fill in the missing spaces on the rows with perimeter 42 and bottom edge 10.

Worksheet 2-1, page 4

30

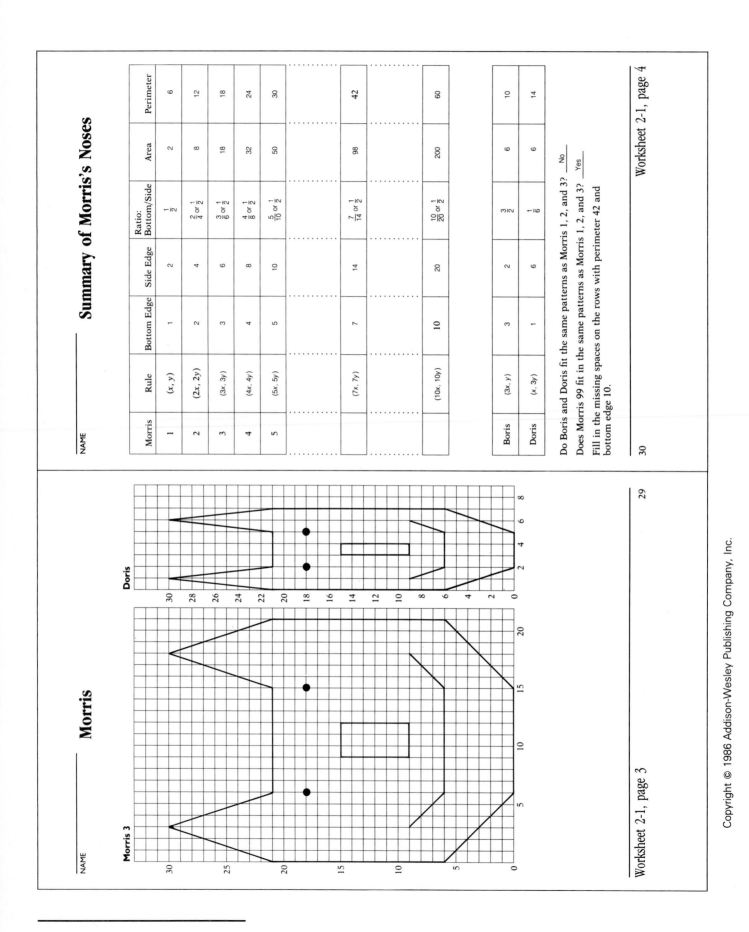

Morris

NAME

Morris 3

Doris

Worksheet 2-1, page 3

29

Similar Rectangles

NAME

Rectangle	Short side a	Long side b	Ratio $\frac{a}{b}$	Area	Perimeter
1	3 units	4 units	$\frac{3}{4}$	12 sq units	14 units
2	6	8	$\frac{6}{8}$ or $\frac{3}{4}$	48	28
3	9	12	$\frac{9}{12}$ or $\frac{3}{4}$	108	42
4	12	16	$\frac{12}{16}$ or $\frac{3}{4}$	192	56
5	15	20	$\frac{15}{20}$ or $\frac{3}{4}$	300	70
6	10	14	$\frac{10}{14}$ or $\frac{5}{7}$	140	48

Which rectangles are similar? 1, 2, 3, 4, 5

Give a rule for testing rectangles to see if they are similar.
Ratio of short side to long side is the same.

44

Worksheet 3-2

Morris 99

NAME

Points	Morris 1 (x, y)	Morris 99 $(2x + 4, 2y + 6)$
Set I		
A	5, 0	14, 6
B	7, 2	18, 10
C	7, 7	18, 20
D	6, 10	16, 26
E	5, 7	14, 20
F	2, 7	8, 20
G	1, 10	6, 26
H	0, 7	4, 20
I	0, 2	4, 10
J	2, 0	8, 6
Connect to A		
Start Over		
Set II	K	1, 3	6, 12
L	2, 2	8, 10
M	5, 2	14, 10
N	6, 3	16, 12
Start Over		
Set III	O	3, 3	10, 12
P	4, 3	12, 12
Q	4, 5	12, 16
R	3, 5	10, 16
Connect to O		
Set IV	S	5, 6 (DOT)	14, 18
Set V	T	2, 6 (DOT)	8, 18

31

Worksheet 2-2

Answers

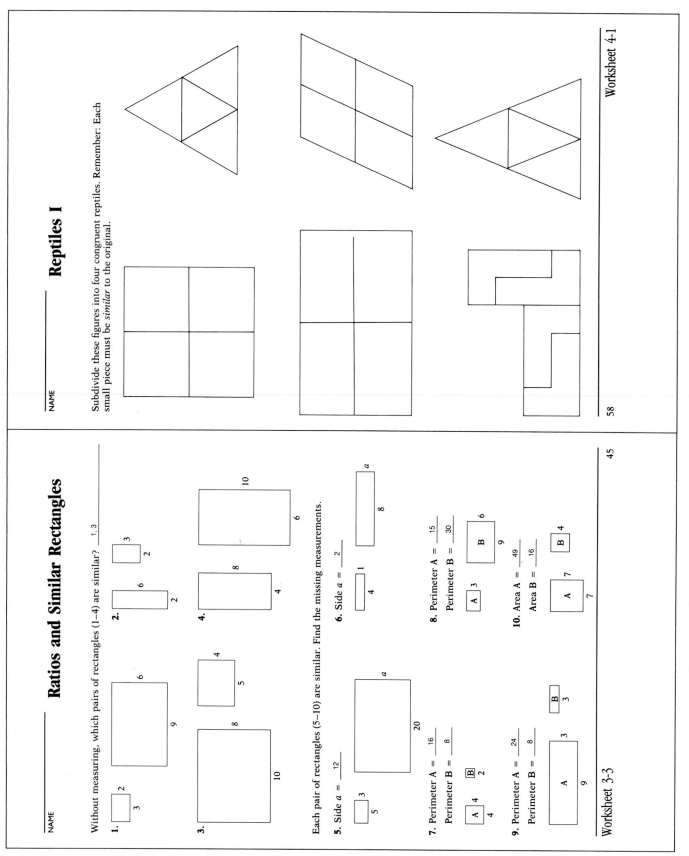

Answers

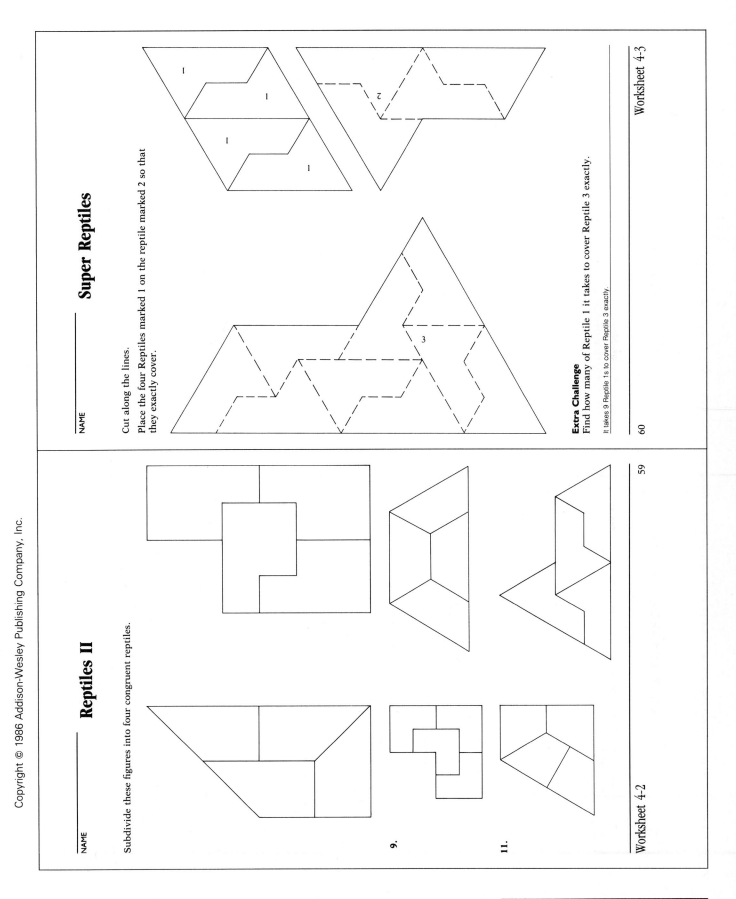

NAME

Reptiles II

Subdivide these figures into four congruent reptiles.

9.

11.

Worksheet 4-2

59

NAME

Super Reptiles

Cut along the lines.

Place the four Reptiles marked 1 on the reptile marked 2 so that they exactly cover.

1

1

1

1

2

3

Extra Challenge
Find how many of Reptile 1 it takes to cover Reptile 3 exactly.

It takes 9 Reptile 1s to cover Reptile 3 exactly.

60

Worksheet 4-3

155

Answers

Chart for Similar Triangles

NAME

Which triangles are similar?

How do similar right triangles grow? Use triangle 5 to find the scale factor for the other similar triangles.

Describe how you can test two right triangles to see if they are similar.

Ratios and corresponding sides are equal.

Triangle	Sides			Ratio $\frac{a}{b}$	Scale Factor	Perimeter	Area
	a	b	c				
1	9	12	15	$\frac{9}{12} = \frac{3}{4}$	3	36 units	54 sq. units
2	15	20	25	$\frac{15}{20} = \frac{3}{4}$	5	60	150
3	18	24	30	$\frac{18}{24} = \frac{3}{4}$	6	72	216
4	10	14	$17\frac{1}{2}$	$\frac{10}{14} = \frac{5}{7}$	X	X	X
5	3	4	5	$\frac{3}{4}$	1	12	6
6	6	8	10	$\frac{6}{8} = \frac{3}{4}$	2	24	24
7	12	16	20	$\frac{12}{16} = \frac{3}{4}$	4	48	96
8	$4\frac{1}{2}$	6	$7\frac{1}{2}$	$\frac{4\frac{1}{2}}{6} = \frac{3}{4}$	$1\frac{1}{2}$	18	$13\frac{1}{2}$

Worksheet 5-1, page 2

Are They Similar?

NAME

For each pair of figures show why they are or are not similar. You should check corresponding angles and ratios of corresponding sides to see that they are or are not equal.

1.

Corresponding angles are equal.
Ratios of sides are equal to $\frac{1}{2}$.
Figures are similar.

2.

Corresponding angles are not equal.
Ratios of sides are equal to $\frac{1}{2}$.
Figures are not similar.

3.

Corresponding angles are equal.
Ratios of sides are not equal.
Figures are not similar.

4.

Corresponding angles are equal.
Ratios of sides are equal to $\frac{1}{3}$.
Figures are similar.

5.

Corresponding angles are equal.
Ratios of sides are not equal.
Figures are not similar.

6. Hint: Use the $\frac{1}{2}$ cm grid over the figures.

Corresponding angles are equal.
Ratios of sides are equal to $\frac{1}{3}$.
Figures are similar.

Worksheet 4-4

Answers

Testing Right Triangles

NAME _____

A. Without measuring, tell which pairs are similar. ____ 1, 3, 4

1. Yes

2. No

3. Yes

4. Yes

B. Each pair of right triangles is similar. Find the missing measurement.

5. Side a = ____ 8

6. Side a = ____ 20

7. Side A = ____ 9

8. Side A = ____ 3

74   Worksheet 5-2

Rectangles

NAME _____

1. Record the data.

Rectangle	A	B	C	D	E	F	G	H
Short Side	20 units	16	10	8	5	4	$2\frac{1}{2}$	2
Long Side	32 units	20	16	10	8	5	4	$2\frac{1}{2}$
Ratio $\frac{\text{short}}{\text{long}}$	$\frac{5}{8}$	$\frac{4}{5}$	$\frac{5}{8}$	$\frac{4}{5}$	$\frac{5}{8}$	$\frac{4}{5}$	$\frac{5}{8}$	$\frac{4}{5}$

2. Record the families of similar rectangles that you find in the chart.

Family I: ____ A C E G

Family II: ____ B D F H

3. What patterns do you see? ____ Ratios alternate $\frac{5}{8}$ then $\frac{4}{5}$

		Family I				Family II	
Rectangles	A	C	E	G	B	D	F	H
Area	640 sq. units	160	40	10	320	80	20	5
Perimeter	104	52	26	13	72	36	18	9

4. How are the perimeters and areas related in each family of similar rectangles?

The areas decrease by a factor of $\frac{1}{4}$ each time. The perimeters decrease by a factor of $\frac{1}{2}$ each time.

5. Describe every different test for similar rectangles that you have studied.

Ratios of corresponding sides are equal. When rectangles are nested in the lower left hand corner, the diagonals coincide.

6. Consider the pair of rectangles A and E. What is the scale factor from A to E? ____ $\frac{1}{4}$ What is the scale factor from E to A? ____ 4

88   Worksheet 6-1, page 2

157

Answers

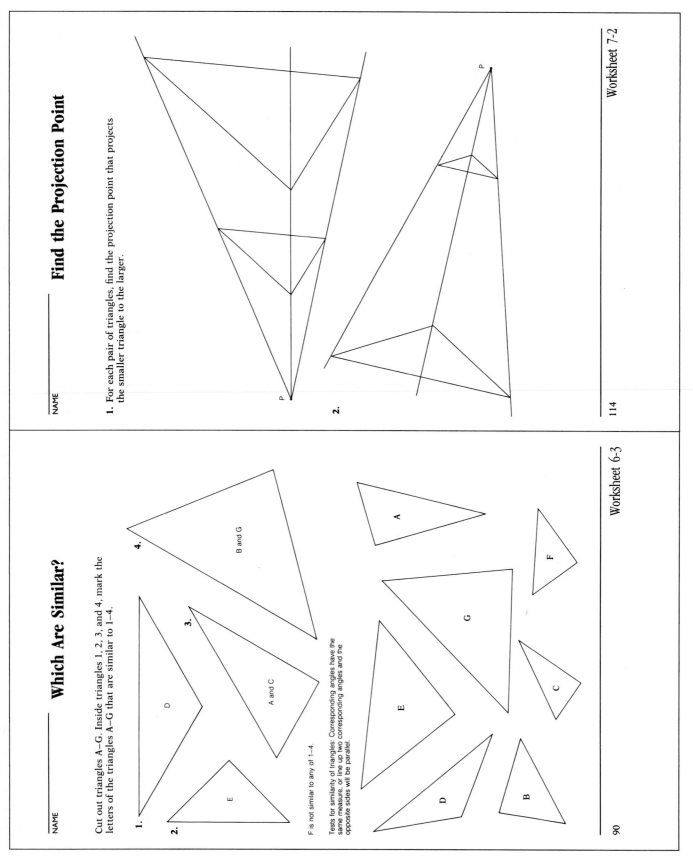

1. For each pair of triangles, find the projection point that projects the smaller triangle to the larger.

P

2.

P

Cut out triangles A–G. Inside triangles 1, 2, 3, and 4, mark the letters of the triangles A–G that are similar to 1–4.

4.

B and G

1.

D

3.

A and C

2.

E

F is not similar to any of 1–4.

Tests for similarity of triangles: Corresponding angles have the same measure, or line up two corresponding angles and the opposite sides will be parallel.

A

F

G

C

E

D

B

Answers

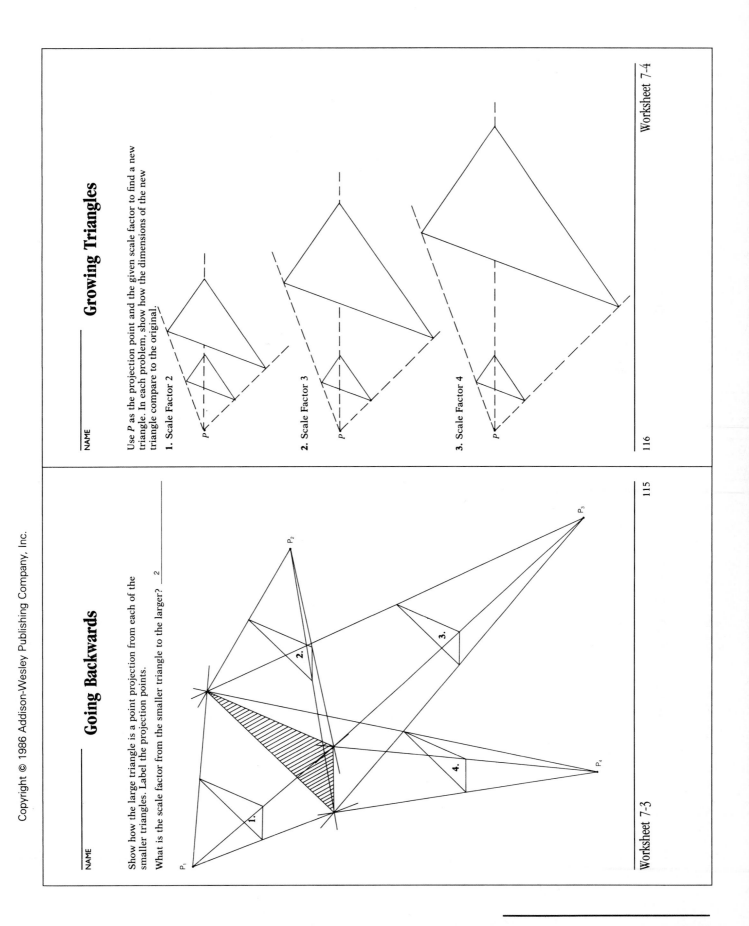

Growing Triangles

NAME _____

Use *P* as the projection point and the given scale factor to find a new triangle. In each problem, show how the dimensions of the new triangle compare to the original.

1. Scale Factor 2

2. Scale Factor 3

3. Scale Factor 4

116

Worksheet 7-4

Going Backwards

NAME _____

Show how the large triangle is a point projection from each of the smaller triangles. Label the projection points.

What is the scale factor from the smaller triangle to the larger? ___2___

Worksheet 7-3

115

159

Answers

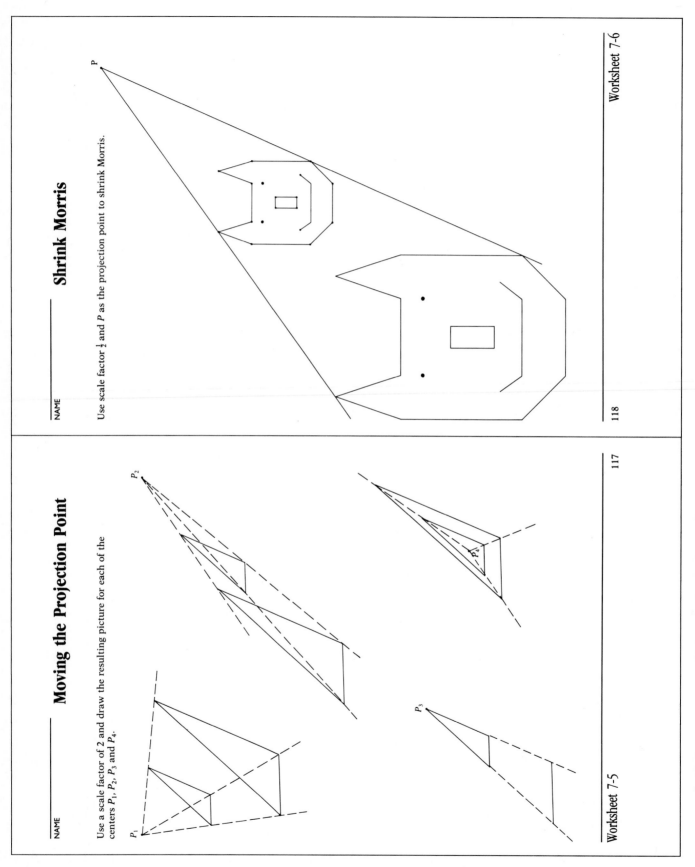

NAME

Moving the Projection Point

Use a scale factor of 2 and draw the resulting picture for each of the centers P_1, P_2, P_3 and P_4.

P_2

P_1

P_3

P_4

Worksheet 7-5

117

NAME

Shrink Morris

Use scale factor $\frac{1}{2}$ and P as the projection point to shrink Morris.

P

118

Worksheet 7-6

Answers

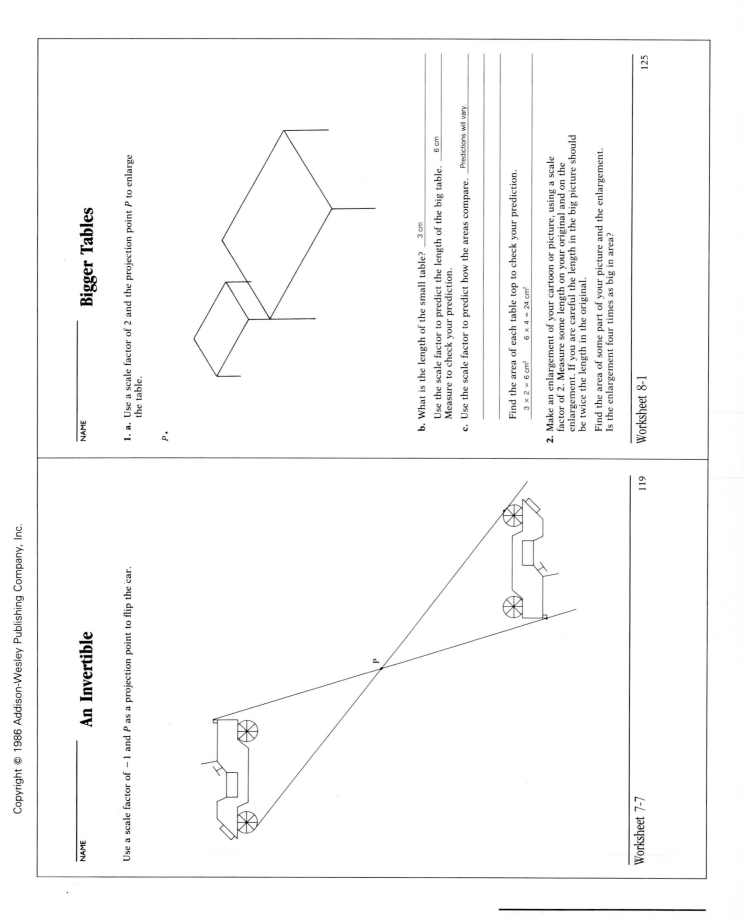

NAME

An Invertible

Use a scale factor of −1 and *P* as a projection point to flip the car.

P

Worksheet 7-7

119

NAME

Bigger Tables

1. a. Use a scale factor of 2 and the projection point *P* to enlarge the table.

P.

b. What is the length of the small table? 3 cm

Use the scale factor to predict the length of the big table. 6 cm
Measure to check your prediction.

c. Use the scale factor to predict how the areas compare. Predictions will vary.

Find the area of each table top to check your prediction.

3 × 2 = 6 cm²   6 × 4 = 24 cm²

2. Make an enlargement of your cartoon or picture, using a scale factor of 2. Measure some length on your original and on the enlargement. If you are careful the length in the big picture should be twice the length in the original.

Find the area of some part of your picture and the enlargement. Is the enlargement four times as big in area?

Worksheet 8-1

125

161

Answers

Application Problems

In each problem, first draw a picture. Then decide whether the problem involves measuring length or area growth.

1. It costs Mrs. Jones $400 to carpet the small room shown. She has another room twice as long and twice as wide. How much will it cost her to carpet the bigger room in the same carpeting? Before you answer, draw a picture of the bigger room. Is the question asked about the growth of length or of area?

$$4 \times 400 = \$1,600$$

Suppose Mrs. Jones wanted to put the same carpet in a room 3 times as long and 3 times as wide. How much would it cost?

$$9 \times \$400 = \$3,600$$

2. Sue is building a model boat. The scale factor from the model boat to the real boat is 20. If the big boat is 40 feet long, how long is the model?

$$40 = 20 \times 2 \qquad \underline{2\ ft}$$

If the model boat has doors that are 2 inches high, how high are the doors on the real ship?

$$\underline{40\ inches}$$

If it takes $\frac{1}{3}$ can of paint to paint the deck of the model boat, how much paint will it take to paint the deck of the real boat?

$$\frac{1}{3} \times (20)^2 = \frac{400}{3} \qquad \underline{133\tfrac{1}{3}\ cans}$$

3. The toy shop has a doll house copy of a real house built on a scale of 1' to 10'. This means that the scale factor from the doll house to the real house is 10. If the picket fence around the doll house is 14 feet long, how long is the picket fence around the real house?

$$P = 14 \times 10 = 140\ ft \qquad \underline{140\ ft}$$

If it takes 200 yards of fabric to make the curtains and bedspreads for the real house, how much fabric is needed for the doll house?

$$\frac{200}{10^2} = 2 \qquad \underline{2\ yd}$$

The toy house has 3 chimneys, how many chimneys does the real house have?

$$\underline{3}$$

Indirect Measurement

1. The picture below shows how Mary used a short tree to find the height of the tall tree.

What triangles are similar? ABE and ACD

What is the height of the tall tree?

$$\frac{?}{30} = \frac{6}{10} = \frac{6 \times 3}{10 \times 3} = \frac{18}{30}$$

$$? = 18\ ft$$

2. José used a stick and shadows to find the height of a tree. What answer should he get?

Stick, 80 cm

Shadow, 200 cm

Shadow, 35 meters or 3500 cm

$$\frac{height}{3500} = \frac{80}{200} = \frac{40}{10} = \frac{4 \times 350}{10 \times 350} = \frac{1400}{3500}$$

$$height = 1400\ cm\ or\ 14\ meters$$

3. Paul and Samantha want to find how wide a natural rock tower is. First they locate a point A from which they can sight the edge of the rock on both sides. From A they draw lines to both edges of the rock and beyond. Then they stake points B and C so that the line from B to C is parallel to the width from D to E that they want to measure. Then they measure the distances, as shown in the diagram. How wide is the rock tower?

$$\frac{?}{30} = \frac{24}{40} = \frac{3}{5} = \frac{3 \times 6}{5 \times 6} = \frac{18}{30}$$

$$? = 18$$

The rock tower is 18 m wide

Answers

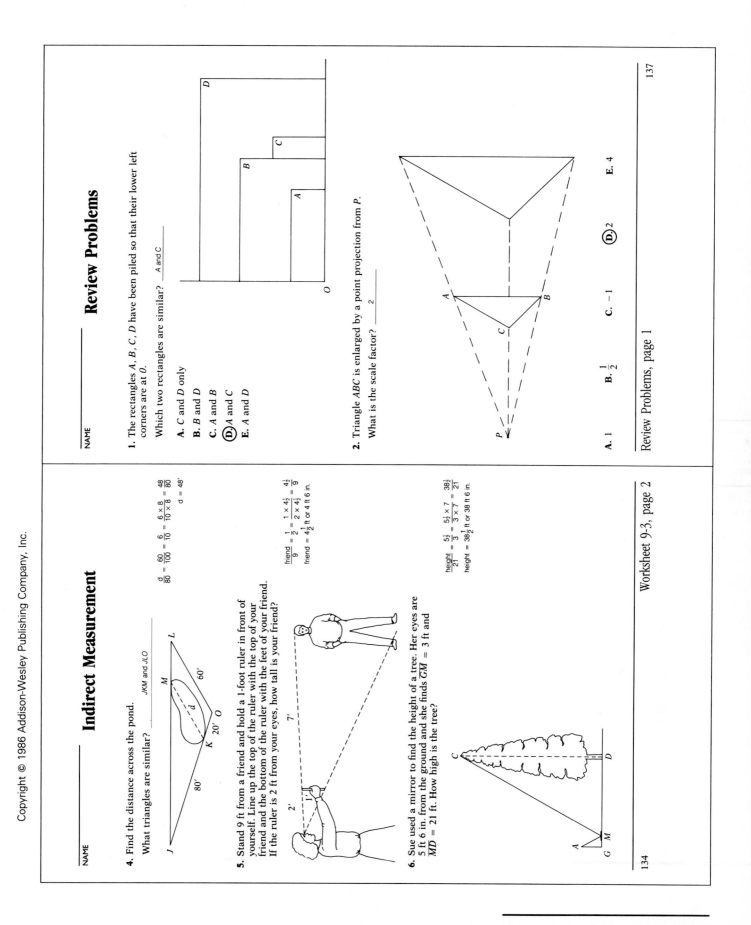

Review Problems

NAME

1. The rectangles A, B, C, D have been piled so that their lower left corners are at O.

Which two rectangles are similar? _____ A and C

- **A.** C and D only
- **B.** B and D
- **C.** A and B
- **Ⓓ.** A and C
- **E.** A and D

2. Triangle ABC is enlarged by a point projection from P.

What is the scale factor? _____ 2

A. 1   **B.** $\frac{1}{2}$ **C.** -1 **Ⓓ.** 2   **E.** 4

Review Problems, page 1

137

Indirect Measurement

NAME

4. Find the distance across the pond.

What triangles are similar? _____ JKM and JLO

$$\frac{d}{80} = \frac{60}{100} = \frac{6}{10} = \frac{6 \times 8}{10 \times 8} = \frac{48}{80}$$
$$d = 48'$$

5. Stand 9 ft from a friend and hold a 1-foot ruler in front of yourself. Line up the top of the ruler with the top of your friend and the bottom of the ruler with the feet of your friend. If the ruler is 2 ft from your eyes, how tall is your friend?

$$\frac{\text{friend}}{9} = \frac{1}{2} = \frac{1 \times 4\frac{1}{2}}{2 \times 4\frac{1}{2}} = \frac{4\frac{1}{2}}{9}$$
$$\text{friend} = 4\frac{1}{2} \text{ ft or 4 ft 6 in.}$$

6. Sue used a mirror to find the height of a tree. Her eyes are 5 ft 6 in. from the ground and she finds $GM = 3$ ft and $MD = 21$ ft. How high is the tree?

$$\frac{\text{height}}{21} = \frac{5\frac{1}{2}}{3} = \frac{5\frac{1}{2} \times 7}{3 \times 7} = \frac{38\frac{1}{2}}{21}$$
$$\text{height} = 38\frac{1}{2} \text{ ft or 38 ft 6 in.}$$

Worksheet 9-3, page 2

134

163

Answers

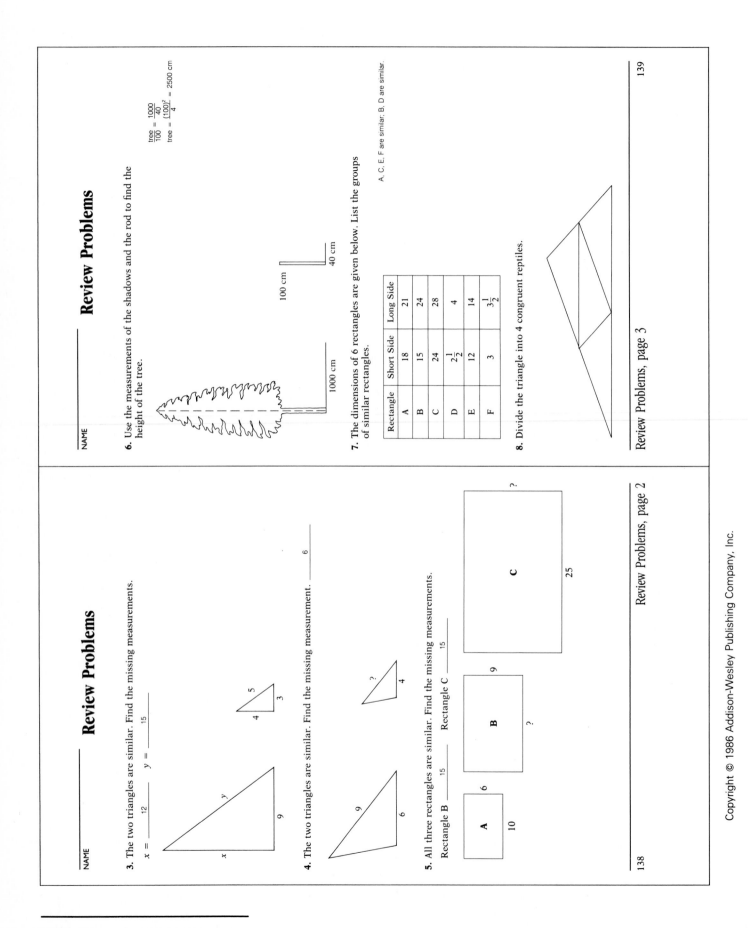

Review Problems

3. The two triangles are similar. Find the missing measurements.

$x =$ ___12___ $y =$ ___15___

4. The two triangles are similar. Find the missing measurement. ___6___

5. All three rectangles are similar. Find the missing measurements.

Rectangle B ___15___   Rectangle C ___15___

138   Review Problems, page 2

Review Problems

6. Use the measurements of the shadows and the rod to find the height of the tree.

$\frac{tree}{100} = \frac{1000}{40}$

$tree = \frac{(100)^2}{4} = 2500$ cm

1000 cm   100 cm

40 cm

7. The dimensions of 6 rectangles are given below. List the groups of similar rectangles.

Rectangle	Short Side	Long Side
A	18	21
B	15	24
C	24	28
D	$2\frac{1}{2}$	4
E	12	14
F	3	$3\frac{1}{2}$

A, C, E, F are similar; B, D are similar.

8. Divide the triangle into 4 congruent reptiles.

Review Problems, page 3   139

Answers

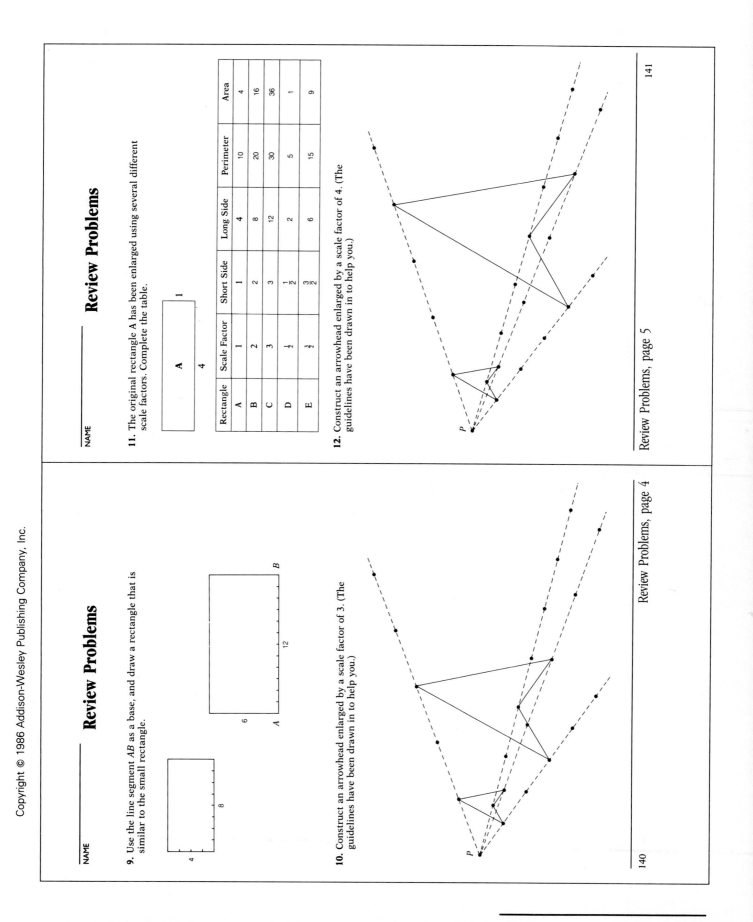

Review Problems

NAME

9. Use the line segment *AB* as a base, and draw a rectangle that is similar to the small rectangle.

8

4

6

12

A

B

10. Construct an arrowhead enlarged by a scale factor of 3. (The guidelines have been drawn in to help you.)

P

Review Problems

NAME

11. The original rectangle A has been enlarged using several different scale factors. Complete the table.

A   4   1

Rectangle	Scale Factor	Short Side	Long Side	Perimeter	Area
A	1	1	4	10	4
B	2	2	8	20	16
C	3	3	12	30	36
D	$\frac{1}{2}$	$\frac{1}{2}$	2	5	1
E	$\frac{3}{2}$	$\frac{3}{2}$	6	15	9

12. Construct an arrowhead enlarged by a scale factor of 4. (The guidelines have been drawn in to help you.)

P

Unit Test Answer Key

1. C		16. C
2. E		17. D
3. E		18. B
4. E		19. B
5. A		20. D
6. C		21. D
7. A		22. D
8. C		23. B
9. E		24. A
10. C		25. A
11. D		
12. B		
13. A		
14. E		
15. E		